Solving the UFO Enigma

How Modern Physics is Revealing the Technology of UFOs

Robert L. Schroeder

Quantum Universe Books / South Carolina

Quantum Universe Books / North Charleston, South Carolina

Printed in the United States of America

Cover photo: Picture taken by J.S. Henrardi on
June 15, 1990 in Wallonia, Belgium.
The photographer has released this picture into
the public domain.

ISBN-10: 1452836817
ISBN-13: 978-1452836812
BISAC: Science/Physics
LCCN: 2011915617

Contents

Contents

Foreword

The thesis of this book is that not only are some UFOs real craft from extraterrestrial civilizations but that scientific progress being made on planet earth will ultimately reveal the technology behind the UFO phenomenon.   My goal is to show how the science of physics has evolved over the last century and how its direction is already pointing toward a potential solution to the UFO enigma.   Experiments now underway at the Large Hadron Collider particle accelerator in Geneva, Switzerland are looking for, among other things, evidence of extra dimensions and the so called Higgs particle.   The results of these experiments I believe will be very important in helping to unravel the puzzle of UFO engineering.

The idea that at least some UFOs are vehicles from extraterrestrial civilizations has been around for a long time.   There are good reasons to believe that this is the case.   Even the generally skeptical study of UFOs sponsored by the U.S. Air Force in the late 1960's known as the Condon Report included a number of sightings that baffled the experts.   More recently the French government in its semi-official COMETA Report concluded that the best explanation for at least some UFOs is that they are of extraterrestrial origin.   Many other researchers have come to the same conclusion.   For the purposes of this book my working assumption is that a subset of all UFO reports does represent real machines from civilizations beyond earth.

But the fascinating question is that if these really are extraterrestrial craft from distant solar systems then how did they get here?   Even the nearest star system to earth, Alpha Centauri, is over four light years from our planet.   Our best technology currently available would still involve a trip taking centuries just to get to Alpha Centauri.   Yet there are reasons to suspect that UFOs reach planet earth fairly easily.   For one thing credible UFO sightings occur quite frequently.   Perhaps even more revealing is the often frivolous behavior of UFOs.   For example there are a number of cases on record where UFOs have toyed with cars, usually on remote roads.   This is not what you would expect if extraterrestrials had spent years or decades traveling to our planet.

So what is the technology behind these alien craft?   Not only is there the question of how they got here but also the puzzle of the often phenomenal performance characteristics of UFOs in earth's atmosphere.   The UFO data by itself has not been sufficient to answer this question. However developments in modern physics combined with selected UFO data is, I believe, already giving us the information needed for a likely solution to UFO technology.   That is the theme of this book.

The book is divided into three sections.   Section I presents the UFO data.   I was careful to include only what I considered to be the more reliable UFO reports.

Section II explains the evolution of modern physics right up to the latest cutting edge theories. To understand where we are now it is important to see how physics has evolved and to be aware of the *unanswered questions* in known physics.   For example a key postulate (assumption) in Einstein's theory of general relativity is the equivalence principle which equates inertial mass to gravitational mass.   Yet modern physics does not know the origin of the force we call inertia. As we shall see the answer to some of these puzzles is central to the mystery of UFO technology.

In the last part of the book, Section III, I review the UFO data, summarize modern physics including recent theories and then propose a solution to the UFO enigma.   I also take a look at how an understanding of UFO physics will likely impact our own civilization in the future.

The book does contain some math.   But don't let that scare you!   It's not that hard.   Math is the language of science so it does help to convey the meaning of the scientific concepts in a way that words alone might not always succeed. However the math isn't necessary to understand the general theme of the book. Readers can if they wish merely scan the math by reading the text interspersed with the math and then reviewing the summaries and conclusions at the end of each chapter. Only 5 of the 22 chapters include any mathematics. I included some math because one of my goals with this book is to take the mystery out of science, particularly physics, and make it accessible to a wide audience.

Some of the math includes calculus but I have an easy explanation of calculus in Appendix A for those who are interested in digging into the equations. Calculus is actually pretty easy! If you've ever driven a car (or bike) at constant speed you can think of that as the first derivative. When you accelerate the car that can be thought of as the second derivative. Well that's pretty much what calculus is all about!

I believe we may be a lot closer to understanding and implementing UFO technology than anyone has so far imagined. If the conjectures presented in this book prove to be correct it will open up whole new worlds, both literally and figuratively, to be discovered and explored. Enjoy!

Robert L. Schroeder

Acknowledgements

My wife and daughter deserve special thanks for their assistance and encouragement while hubby/dad wrote this book over the last four years.   Since I'm a terrible typist my wife Carol's typing of a good chunk of the manuscript went a long ways in saving me from many hours of hunt and peck.   Daughter Debbie kept me up to date on computer graphics issues.   It's hard to teach an old dad new tricks!

Our good family friends, Bob and Jane Morrissey, provided some valuable assistance in several areas including suggestions on what topics should get special emphasis in the book.

The unsung heroes of UFO studies are the numerous researchers and field investigators who for many years have diligently and tirelessly gathered the data and tried to make sense of this perplexing phenomenon.   Some of these men and women work independently while others have worked for organizations like NICAP, CUFOS, NARCAP, APRO, NUFORC, MUFON, FUFOR, etc.   You will find the names of many of these great people throughout the book.   They come from assorted backgrounds including engineers, scientists, housewives, students, journalists, medical doctors, astronauts, authors and so on.   It should be mentioned that they have often braved the skepticism and ridicule of their fellow citizens yet have persevered nonetheless.   I am confident that the time will come when they will receive the recognition that they deserve.   Their efforts were essential to the completion of this book.

The progress of modern physics, along with crucial UFO data, was instrumental in motivating me to write this book.   Countless scientists, engineers and technicians have contributed to the amazing success of contemporary physics as well as to the exciting cutting edge theories now facing experimental test.   The names of many of these intrepid individuals and their important accomplishments are included in Section II and elsewhere in this book.

Last but not least I have to thank my publisher for all the help they gave this novice writer in getting this book published.   Author Larry Zafran, who owns the website "MathWithLarry.com" was also a tremendous help with technical issues in this publishing process.

Section I

The UFO Evidence

My Introduction to the UFO Phenomenon
Chapter 1 in Section I

My interest in UFOs goes back to a fall afternoon in 1957 when, as a 12 year old boy, I observed two unusual objects at high altitude passing over my family's house in Teaneck, New Jersey. Teaneck is a small town located several miles west of Manhattan. The objects were apparently heading toward New York City.

The first object appeared to be a cylinder about the shape of a cigarette and traveling in the direction of its long axis. This was followed by a smaller circular object which was behind the cigarette shaped craft by about two of the cylinders lengths. The circular object was no wider than the width of the cigarette shaped object. Both objects were a metallic gold color against a clear (about 5 PM) blue sky and both followed exactly the same flight path in tandem. These are illustrated below:

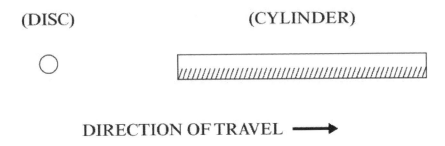

(DISC)   (CYLINDER)

DIRECTION OF TRAVEL ➝

(Objects seen by author in the fall of 1957 over northern New Jersey)

The objects gave the impression of being at high altitude and much larger than a commercial airliner. My first thought was that I might be observing Sputnik, the new Russian satellite, and the booster rocket that put it into orbit. But I quickly dismissed this idea since it seemed unlikely that the booster rocket would have remained in orbit along with the satellite. Also puzzling was why the satellite would be *behind* the booster rocket! Furthermore there was no way I could have seen, with the unaided eye, the beach ball sized Sputnik satellite which I knew was over 100 miles above the earth's surface.

My next thought was that I was observing the new Boeing 707 jetliner and that the wings were tilted slightly such that they were not visible against the blue sky background. I had noticed this effect sometimes with high flying aircraft. But even that explanation fell apart since I knew the 707 was tapered at both the nose and the tail, whereas the cylindrical object was clearly blunt or squared off on both ends.

Finally it seemed to move far too fast to be a blimp. I quickly concluded that I was observing something very unusual. I ran inside the house to tell my brothers but by the time I dragged one of them outside the object had disappeared. Although I come from a family of scientists and engineers and am "hard headed" about things, I sensed that I had witnessed something quite extraordinary.   I wondered if I had observed one of the UFOs that had been reported in the news.

In subsequent years I learned much more about the UFO phenomenon through widespread reading and began to notice intriguing similarities among the more credible reports. For example I read a number of reports that described huge cylindrical UFOs that appeared to be carriers for the more typical disk shaped objects.   The disk shaped craft probably being stacked like plates within the large cylindrical UFO.   I realized that the object I had seen in 1957 exactly matched these descriptions and that the circular object trailing behind was most likely a disk seen from below.

And if that were not enough to convince me of the reality of what I had seen, more circumstantial evidence was to follow.   In my extensive readings I also learned that there are periods of increased UFO activity that are referred to as UFO waves.   One of the most intense periods was in the fall of 1957, exactly when I had observed the unusual objects over my New Jersey home.

Finally I came across a comment by the late Dr. J. Allen Hynek, the Northwestern University astronomer and one time advisor to the Air Force's Project Bluebook.   Dr. Hynek observed that most UFO witnesses go through a process he called the "escalation of hypotheses" whereby the witness attempts to explain his sighting in terms of known phenomenon.   This was exactly the process I had followed.

In the summer of 1958 I borrowed a book from the Teaneck, N.J. public library titled "Flying Saucers From Outer Space" by Major Donald Keyhoe and took it with me to our family's summer house on Cape Cod.   Although Keyhoe's book was a bit sensationalized I realized immediately that many credible witnesses such as pilots, radar operators and scientists were observing truly incredible things.   The descriptions clearly pointed in the direction of intelligently controlled craft, most likely from beyond our solar system.   The thought was electrifying.   Many additional well documented and credible reports since the 1950's have convinced me that the extraterrestrial hypothesis is the most likely explanation for many reported UFOs.

In the intervening years I have read countless books and articles on the UFO phenomenon.   In more recent years there have also been numerous documentaries on TV dealing with UFOs Much of the information on the subject has been quite good, some average, and a certain amount not quite so good to put it charitably.   Nevertheless there is a large body of anecdotal data on the topic and some "hard" data such as radar tapes, movie films, still pictures and video recordings.

Coming from a math and science background I have been extremely curious about the possible engineering behind the UFO phenomenon.   Over the years I have studied modern physics as much as time allowed with family responsibilities and while pursuing a busy career in the computer industry.   In the last few years I recognized that some of the latest theories in physics, combined with some interesting conjectures and selected UFO data, offered a potential solution to UFO technology.   This is the purpose of this book.

In the next four chapters of Section I we'll take a look at what I believe is the more credible UFO data.   In Section II we'll look at the evolution of modern physics right up to the most current theories.   Finally in Section III I will present a solution to the UFO enigma based on merging the UFO data with modern physics including exciting developments in both experimental and theoretical physics.

The Classic UFO Cases
Chapter 2 in Section I

Introduction

For this classic UFO sightings chapter I have selected 12 cases which I think are exceptional from the point of view of witness credibility, extensiveness of investigation and how they clearly demonstrate the nature of the UFO phenomenon. Over the years I have read many interesting books and articles on the subject of UFOs. While much of the content of this literature has been quite good, I nevertheless have often found myself disappointed to see good UFO reports from credible witnesses mixed in with dubious reports, some of which give every indication of having been fabricated. This has undoubtedly played a significant role in discouraging serious scientific research of the topic. So for this reason I have tried my best to select only what I believe are the more substantial cases throughout Section I. Despite this cautious approach I cannot absolutely guarantee the validity of any case, only that I have used my years of experience to carefully filter the data.

The classic cases range in time from the 1930's up to the 1980's. Most are multiple witness cases but some are well investigated single witness cases. Many readers may already be familiar with some or all of the cases but I would still urge them to review these reports since there are certain salient features in each case that are important to the general theme of this book. Other readers who are newly exposed to the subject matter are likely to find themselves surprised at the caliber and quality of these UFO cases.

I will use Dr. Hynek's UFO report classification system in this chapter and throughout the book. Dr. Hynek classifies UFO reports into two general categories; namely UFOs seen at a distance, and Close Encounter cases where the UFO is observed within 500 feet of the witnesses. Each of these categories is further broken down into subcategories which are shown below:

Distant UFOs

Daylight Discs/Lights - Generally disc or oval shaped objects observed in daylight, sometimes illuminated. The shape is loosely defined here and can include other configurations as well.

Nocturnal Lights - Strangely behaving lights in the night sky that defy conventional explanation.

Radar Visual - A visual UFO observation that is accompanied by a radar sighting.

Close Encounters

Close Encounters of the First Kind (CE-I) - Here the UFO is seen at close range but there is no interaction with the environment (other than possible trauma on the part of the witness).

Close Encounters of the Second Kind (CE-II) - Similar to a CE-I except that physical effects on both animate and inanimate matter are noted. For example there is often damage to vegetation or animals are frightened. Cars are reported to have had their engines cease to function and lights and radio falter or fail altogether. Usually the vehicles return to normal operation after the UFO has departed the scene.

Close Encounters of the Third Kind (CE-III) - In these cases the presence of occupants is reported within the UFO or in its vicinity. This category, however, does not include "contactee" cases.

Following are the classic cases:

Classic Case #: 1
Report Type: Daylight Discs
Location: New England
Witnesses: Family
Date: Spring of 1936 or 1937

Many people dismiss the UFO phenomenon as a post World War II case of "war nerves". But in actuality there are a number of well documented sightings of UFOs that occurred well before World War II. This next sighting was described in Dr. J. Allen Hynek's excellent book "The Hynek UFO Report".

In the early spring of 1936 or 1937 one of the principal witnesses in this sighting, a 13 year old boy, was on a family car outing on a Sunday afternoon. While craning his head backward to eat a taffy apple he observed about a dozen disc shaped objects through the car's rear window in the sky above. Some of the discs cavorted in the sky in quick, jerky movements with sudden stops and starts but no discernible acceleration or deceleration. Other discs simply hovered in place. After ten minutes of this activity the discs suddenly gathered together, as if on cue, in one location under the clouds and stacked up one above the other like plates. The entire formation then ascended vertically right through the clouds and was lost to sight.

All the discs were identical in size and color.

Classic Case#: 2
Report Type: CE-III
Location: Germany
Witnesses: Father and daughter
Date: July 1, 1952

If UFOs are of extraterrestrial origin, as I believe, then what are the occupants likely to look like? While science fiction writers and some scientists predict various bizarre shapes and forms for intelligent creatures from outer space I suspect that they are more likely to resemble ourselves. My guess is that on most planets primate type creatures living in three dimensional forest canopies are more likely to evolve bigger brains than other creatures.

It is certainly exciting to imagine life forms from outer space to be based on silicon or with three eyes and six legs but I think the reality will be more mundane though still fascinating. Needless to say stories of encounters with aliens are fertile ground for dedicated hoaxers. For this reason I carefully review UFO sightings involving occupants and accept only a selected few based on the caliber of the witnesses and various other aspects of the case.

In this Classic UFO Cases chapter I have included three encounters involving occupants. This first one was reported in East Germany in 1952 by a former Wehrmacht major and his daughter. A UFO wave of worldwide proportions occurred in 1952 which helps authenticate this case.

Following is an article on this sighting that appeared in a Greek newspaper on July 9, 1952 which includes eyewitness testimony by the principal witness:

BERLIN- Furnished with sworn testimony of an eyewitness, Oscar Linke, a forty-eight-year-old German, and former mayor of Gleimershausen, West Berlin, intelligence officers have begun investigating a most unusual "flying saucer" story. According to this story, an object "resembling a huge frying pan" and having a diameter of about fifteen meters landed in a forest clearing in the Soviet zone of Germany.

Linke recently escaped from the Soviet zone along with his wife and six children.

Linke and his eleven-year-old daughter Gabriella made the following sworn statement last week before a judge:

"While I was returning to my home with Gabriella, the tire of my motorcycle blew out near the town of Hasselbacht. While we were walking along towards Hasselbacht, Gabriella pointed out something that lay at a distance of about a hundred and forty meters away from us. Since it was twilight, I thought that she was pointing at a young deer.

"I left my motorcycle near a tree and walked towards the spot which Gabriella had pointed out. When, however, I reached a spot about fifty-five meters from the object, I realized that my first impression had been wrong. What I had seen were two men who were not more than forty meters away from me. They seemed dressed in some shiny metallic clothing. They were stooped over and were looking at something lying on the ground.

"I approached until I was only about ten meters from them. I looked over a small fence and then I noticed a large object whose diameter I estimate to be within thirteen to fifteen meters. It looked like a huge frying pan. There were two rows of holes along the periphery, about thirty centimeters in circumference. The space between the two holes was about 0.45 m. On top of this metal object was a black conical tower about three meters high. At that moment, my daughter, who had remained a short distance behind me, called me. The two men must have heard my daughter's voice because they immediately jumped on the conical tower and disappeared inside. I had previously noted that one of the men had a lamp on the front part of his body which lit up at regular intervals. Now, the side of the object on which the holes had opened began to glitter. Its color seemed green but later turned to red. At the same time, I began to hear a slight hum. When the brightness and the hum increased, the conical tower began to slide down into the center of the object. The whole object then began to rise slowly from the ground and rotate like a top.

"It seemed to me as if it were supported by the cylindrical plant which had gone down from the top of the object through the center and had now appeared on the bottom of the object. The object, surrounded by a ring of flames, was now a certain number of feet above the ground.

"I then noted that the whole object had risen slowly off the ground. The cylinder on which it was supported had now disappeared within its center and had reappeared on the top of the object.

The rate of climb had now become greater. At the same time, my daughter and I heard a whistling sound similar to that heard when a bomb falls.

"The object rose to a horizontal position, turned towards a neighboring town, and then, gaining altitude, it disappeared over the heights and forests in the direction of Stockhelm."

Many other persons who live in the same area as Linke related that they saw an object which they thought to be a comet (meteor). A shepherd stated that he thought he was looking as a comet moving away at low altitude from the height on which Linke stood.

After submitting his testimony to the judge, Linke made the following statement: "I would have thought that both my daughter and I were dreaming if it were not for the following elements involved: when the object had disappeared, I went to the place where it had been. I found a circular opening in the ground and it was quite evident that it was freshly dug. It was exactly the same shape as the conical tower. I was then convinced that I was not dreaming." Linke continued: "I had never heard of the term flying saucers before I escaped the Soviet zone into western Berlin. When I saw this object, I immediately thought that it was a new Soviet military machine. I confess that I was seized with fright because the Soviets do not want anyone to know about their work. Many persons have been restricted in their movements for many years in East Germany because they knew too much!"

Classic Case #: 3
Report Type: CE-II
Location: Levelland, Texas
Witnesses: Police officers and civilians
Date: November 2, 1957

Something was definitely happening in the United States in the fall of 1957 involving UFOs My own sighting in New Jersey was around this time and there was a veritable tidal wave of sightings elsewhere in the nation. Among the most interesting cases were a flurry of sightings in the vicinity of Levelland, Texas involving numerous close encounters including many instances where car engines and lights failed while in the presence of the unknown objects.

Around 11 PM on November 2, 1957 the on duty officer, A. J. Fowler, at the Levelland police station received a call from a terrified motorist reporting a strange incident. Pedro Saucedo had been driving along in his truck some four miles west of Levelland when he and his passenger, Joe Salaz, noticed a brilliantly lighted object approach their vehicle. As the object passed over them the truck lights failed and the engine stopped. When the object moved away the lights came back on by themselves and Saucedo had no trouble starting the engine.

Officer Fowler was puzzled but wondered if maybe the men had been drinking. He would soon change his mind.

An hour later Officer Fowler received another call from a motorist who had come upon a brilliantly lit egg shaped object sitting on the road directly in front of him. The incident occurred about four miles east of Levelland. The object was around 200 feet long and so bright that it illuminated the entire area. As he got nearer to the object the lights and engine of his car both failed. The motorist then decided to get out of his car but as he did so the object rose up several hundred feet into the air and its brilliant lights went out. He was then able to start his car.

Shortly after this report Fowler received another call, this time from a motorist eleven miles north of Levelland. This motorist described an experience similar to the other motorists. His lights and engine failed as he approached a brilliantly lit object sitting on the road. But everything returned to normal as soon as the object departed.

Several more motorists called in with similar stories. This prompted Fowler to notify other on duty police to begin investigating the strange incidents. Later that night several police officers did indeed see very unusual lights although not quite as close as those experienced by the other motorists.

Of possible significance in providing clues as to how UFOs function were color changes noted by some observers. When the object was on the ground its color was a brilliant blue-green, but when the object was descending or rising it was always a reddish-orange color.

By the time the night was over Officer Fowler had recorded 15 phone calls that were directly related to the UFO sightings and encounters.

Classic Case#: 4
Report Type: CE-II
Location: Nevada
Witness: Air Force first lieutenant
Date: November 23, 1957

This sighting by an Air Force first lieutenant is one of the most interesting that I've ever come across in many years of reading UFO reports. The lieutenant's close observation of four UFOs confirms many of the structural features and other characteristics that have been observed at greater distances in other sightings.

There were certain observations the witness made in this case that convinced me that this report is entirely authentic. Additionally this sighting has a high credibility since a military officer is unlikely to fabricate a UFO story as this would certainly jeopardize his career prospects should it be revealed as a hoax. Furthermore, the fall of 1957 was witness to one of the most intense UFO waves the United States has ever experienced. My own sighting of a UFO occurred in the fall of 1957 as well. All of this lends considerable credibility to the Air Force lieutenant's sighting which took place during exactly this same period. Here is the lieutenant's report as it appears in Project Blue Book files:

"Source was returning to Newcastle County Airport, Delaware, after completion of USAF Advanced Survival School, Stead AFB on 23 Nov 1957 in his automobile. At about 0610, he was approximately thirty miles west of Tonopah, Nevada, traveling towards Las Vegas at about eighty MPH when the engine of his car suddenly stopped. Attempts to restart the engine were unsuccessful and Source got out of his car to investigate the trouble. Outside the car he heard a steady high-pitched whining noise which drew his attention to four disc-shaped objects that were sitting on the ground about three hundred to four hundred yards to the right of the highway. These objects were totally unlike anything he had ever seen, and he attempted to get closer for a better look at them. He walked for several minutes until he was within approximately fifty feet from the nearest object. The objects were identical and about fifty feet in diameter. They were disc-shaped, emitting their own source of light causing them to glow brightly. They were equipped with a transparent dome in the center of the top which was obviously not of the same material as the rest of the craft. The entire body of the objects emitted the light. They did not appear to be dark on the underside. They were equipped with three landing gears, each that appeared hemispherical in shape about two feet in diameter and of some dark material. The source estimated the height of the objects from the ground level to the top of the dome to be about ten to fifteen feet. The objects were equipped with a ring around the outside which was darker than the rest of the craft and was apparently rotating. When the Source got to within fifty feet of the nearest object, the hum which had been steady in the air ever since he had first observed the objects, increased in pitch to a degree where it almost hurt his ears, and the objects lifted off the ground. The protruding gears were retracted almost instantly after takeoff; the objects rose about fifty feet into the air and proceeded slowly (about ten MPH) to the north across the highway, contoured over some small hills about a half mile away and disappeared behind those hills. As the objects passed directly over the Source, he observed no evidence of any smoke, exhaust, trail, heat, disturbance to the ground, or terrain or any visible outline of landing gear doors or any other outlines or openings in the bottom. The total time of sighting lasted about twenty minutes. After the objects disappeared, Source examined the place where he had first seen them. There was no evidence that any heat had been present or that the ground had been disturbed in any other way than several very small impressions in the sand where the landing gear had obviously rested. Impressions were very shallow and bowl-shaped, triangular in pattern (an equally sided triangle). Source did not measure the distance between the impressions but estimated it to be about eight to ten feet. After his investigation of the impressions, Source returned to his car and the engine started immediately and ran perfectly. The car Source was driving was a 1956 Chevrolet and he did not experience trouble of similar nature before or after the incident. At the time of sighting, Source had driven from Reno, Nevada, to point of sighting during the night and had slept for about two hours in his car between 2400 and 0200 that same day. Source had had no intoxicants or sleep-retarding drugs. He described his physical condition at the time of sighting as excellent. After the sighting, Source proceeded to Indian Springs AFB, Nevada, where he reported the sighting to the base Security Officer.

"The times of day referred to above are given in Pacific Standard Time. At the time of the sighting, it was daylight although the sun was still behind the mountains. The sun was about to rise in front of the Source. There were no stars or moonlight. There was no overcast. The weather was dry and rather cold and there was no wind. There were no other witnesses to the observation to the best of Source's knowledge."

This sighting by a credible witness gives us some important information about these machines. For one thing the entire body of each UFO was emitting its own source of light. It seems likely that this glow or radiation has something to do with the 'engine' that powers these craft. In many other more distant sightings, especially at night, this all encompassing glow is also observed. The circular shape of the body of these objects must also be significant since it is reported frequently by other witnesses. Another interesting feature is the rotation of a ring or section of the outer circumference of the UFO. The transparent dome on the top clearly suggests that this is where the crew quarters are located. This domelike structure on the top of disk shaped UFOs has also been reported in numerous other sightings.

The witness provided several sketches of the objects which are reproduced below.

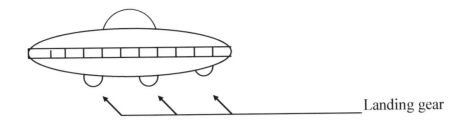

Landing gear

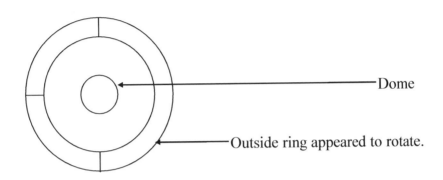

Dome

Outside ring appeared to rotate.

Classic Case#: 5
Report Type: CE-II
Location: Loch Raven Dam, Maryland
Witnesses: Two adults
Date: October 26, 1958

The Loch Raven Dam UFO encounter is an excellent example of the physical effects UFOs apparently produce, either deliberately or inadvertently. Even Project Blue Book, with its generally sloppy analyses, could not come up with an explanation for this classic case so it is listed in the Air Force files as an unknown.

Alvin Cohen and Phil Small on the night of October 26, 1958 were approaching the Loch Raven Dam bridge in Maryland in their car when they noticed a glowing egg shaped object hovering over the steel girdered bridge. The object was motionless in the air and about 100 to 150 feet above the bridge superstructure. When they got to within about 80 feet of the object both the engine and lights of the car failed. Small, the driver, was unable to restart the engine. Apprehensive of the strange craft both men got out and retreated behind the car for what little protection it might provide. In that position they watched the unknown object for some 30 to 40 seconds. Then the object emitted a blinding light accompanied by a loud explosion and rose swiftly into the night sky, disappearing in five to ten seconds.

When the object dramatically increased in brightness both men had felt heat or a burning sensation on their faces. Their faces were flushed as if they had been subjected to a mild sunburn but the effect faded in a few days.

While the object was stationary over the bridge they were able to estimate its width as about 100 feet and that it resembled a flattened egg. But when it ascended vertically into the sky its increased brightness obliterated its shape causing the edges to become diffused.

At a restaurant located one mile north of the bridge some people had heard the explosion and its echo off some nearby cliffs from the object.

Classic Case #: 6
Report Type: CE III
Location: Papua, New Guinea
Witnesses: Reverend Bruce Gill and others at an Anglican mission
Date: June 26-28, 1959

Close encounters with UFOs almost always occur in rural areas and the Papua, New Guinea sightings were no exception to this rule. It is almost as if the operators of UFOs are less concerned about being observed in more remote locations than around cities or more densely populated areas.

In June and July of 1959 there were dozens of sightings of UFOs in Papua, New Guinea But without doubt the most spectacular group of sightings was those made by the Reverend Bruce Gill and 37 other witnesses at an Anglican mission in Boianai, Papua, New Guinea.

On the early evening of June 26, Reverend Gill noticed a bright object just above Venus in the sky. Somewhat later the object descended, coming considerably closer to the mission. Reverend Gill alerted others including members of the staff to come and take a look at the unusual craft. They could see that the object was circular with a wide base and four landing gear extending down from the bottom. Occasionally a thin electric blue spotlight beamed upward at an angle.

What happened next makes this sighting one of the most memorable on record. While watching the hovering UFO they were startled to see movement on the upper surface of the craft. But they were soon able to make out human-like figures moving about on the upper deck as if engaged in some kind of activity. At first there was one or two figures, then three and finally four glowing 'men' appearing on the deck. They could be seen only from about the waist up. The men, or at least some of them, were visible for 15 minutes before they reentered the craft. Clouds gradually covered the entire sky and after a short while the UFO ascended above the clouds.

About an hour later the UFO (or another one) reappeared but this time accompanied by a number of smaller UFOs that cavorted up and down through the cloud layer as if simply for the sheer enjoyment of it. The cloud layer was at approximately 2,000 feet as measured against a nearby mountainside. The craft at various times were well below the cloud layer and theyclearly illuminated the base of the clouds.

The following evening the UFOs appeared again. Around 6 P.M., when it was still light, a large UFO descended to within 300 to 500 feet of the ground. Again the four men appeared on the deck of the hovering UFO but this time one of them seemed to be leaning on the 'railing' looking down at the people in the mission. Thinking the craft might be preparing to land, Reverend Gill waved at the man looking down on them. To everyone's surprise the 'man' waved back with one hand over his head. To be certain he was really responding to the people in the mission one of

Reverend Gills companions waved both hands above his head and, in obvious response, the figure on the UFO waved both hands above his head! Other members of the mission began waving at the men on the UFO and in short order all four men on the craft's deck were waving back at the people on the ground. Finally the figures on the UFO reentered the craft and sometime later the UFO receded in the distance.

The following night the UFOs appeared over the mission again but this time at a higher altitude and no figures could be made out on the upper decks.

This particular sighting provides some of the strongest anecdotal evidence available for the extraterrestrial hypothesis. It is extremely unlikely that an Anglican minister would fabricate such an elaborate story and in collusion with so many other witnesses. The numerous other sightings that occurred throughout Papua, New Guinea around this time period make this one of the most compelling close encounters on record.

Classic Case #: 7
Report Type: CE-III
Location: Socorro, New Mexico
Witnesses: Police Officer
Date: April 24, 1964

This sighting is one of the classic cases involving a landed UFO with occupants. Although it involved a single witness, that person was a police officer which adds considerable weight to the sighting.

Another feature of the case that supports the veracity of the witness is the configuration of the landing gear impressions on the ground relative to the location of a burn mark also found at the site. The landing gear impressions were spaced unevenly on the undulating ground of the landing site. But a theorem in geometry states that if the diagonals of a quadrilateral intersect at right angles, then the midpoints of the sides of the quadrilateral lie on the circumference of a circle. The burn mark was found to be virtually at the center of that circle. Since under certain conditions the center of gravity of the landed craft that the officer observed would have been directly over the center of the circle then the configuration of the landing gear impressions becomes quite important. The implication is that the craft had a self-leveling landing gear system which further increases the probability that a real machine of some type landed at Socorro, New Mexico on April 24, 1964.

The witness, Officer Lonnie Zamora, compiled a complete report within a few hours of the sighting. Portions of his statement are included below:

"About 5:45 P.M., April 24, 1964, while in Socorro Two police car, I started to chase a car due south from west side of the courthouse. Car was apparently speeding and was about three blocks in front. At point on Old Rodeo Street, near George Murillo residence, the chased car was going ahead towards the rodeo grounds. Car chased was a new black Chevrolet...

"At this time I heard a roar and saw a flame in the sky to the southwest some distance away-possibly one-half mile or a mile. Came to mind that a dynamite shack in that area had blown up, decided to leave chased car go. Flame was bluish and sort of orange, too. Could not tell size of flame. Sort of motionless flame. Slowly descending. Was still driving car and could not pay too much attention to the flame. It was a narrow type of flame. It was like a "streamed down"-a funnel type-narrower at the top than at the bottom. Flame possibly three degrees or so in width-not wide.

"Flame was about twice as wide at bottom than top, and about four times as high as top was wide. Did not notice any object at top, did not notice if top of flame was level. Sun was to west and did not help glasses. Could not see bottom of flame because it was behind the hill.

"No smoke noted. Noted some "commotion" at bottom-dust? Possibly from windy day-wind was blowing hard. Clear, sunny sky otherwise-just a few clouds scattered over area.

"Noise was a roar, not a blast. Not like a jet. Changed from high frequency to low frequency and then stopped. Roar lasted possibly ten seconds, I was going towards it at the time on a rough gravel road. Saw flames same color as best I can recall. Sound distinctly from high to low until disappeared. Windows both were down. No other spectators noted-no traffic except the car in front. Car in front might have heard it but possibly did not see it because car in front was too close to hill in front to see flame.

"After the roar and flame, did not note anything while going up the somewhat steep, rough hill-had to back up and try again two more times. Got up about halfway first time, wheels started skidding (roar still going on), had to back down and try again before made the hill. Hill about sixty feet long, fairly steep and with loose gravel and rock. While beginning third time, noise and flame not noted.

"After got to top, traveled slowly on the gravel road westardly. Noted nothing for a while-for possibly ten or fifteen seconds. Went slow looking around for the shack-did not recall exactly where the dynamite shack was.

"Suddenly noted a shiny type object to south about one hundred and fifty to two hundred yards. It was off the road. My green sunglasses over prescription. At first glance, stopped. It looked at first like a car turned upside down. Thought some kids might have turned it over. Saw two people in white coveralls very close to object. One of these persons seemed to turn and look straight at my car and seemed startled-seemed to quickly jump somewhat.

"At this time I began moving my car towards them quickly with the idea to help. Had stopped only about a couple of seconds. Objects were like aluminin-was whitish against the mesa background, not chrome. Seemed like oval in shape and I, at first glance, took it to be an overturned white car. Car appeared turned up like standing on radiator or trunk, at this first glance.

"The only time I saw these two persons was when I had stopped for possibly two seconds or so, to glance at the object. I don't recall noting any particular shape or possibly any hats or headgear. Those persons appeared normal in shape-but possibly they were small adults or large kids. Then paid attention to road while driving. Radioed to sheriff's office , "Socorro Two to Socorro. Possible 10-40 [accident]. I'll be 10-6 [busy]." Out of car, checking the car down in the arroyo.

"Stopped car, was still talking on radio, started to get out; mike fell down, reached back to pick up mike then replaced mike in slot and got out of car. Hardly turned around from car when heard roar (was not exactly a blast), very loud roar-at that close, it was real loud. Not like a jet-know what jets sound like. Started low frequency quickly, then rose in frequency (towards higher tone) and in loudness-loud to very loud. At same time as roar, saw flame Flame was under the object. Object was starting to go straight up-slowly up. Object slowly rose straight up. Flame was light blue and at bottom was sort of orange-colored. From this angle, saw what might be the side of object (not end as first noted). Difficult to describe flame. Thought, from roar, it might blow up. Flame might have come from underside of object, at middle, possibly a four feet area-very rough guess. Cannot describe flames further except blue and orange. No smoke except dust in immediate area.

"Object was oval in shape. It was smooth-no windows or doors. As roar started, it was still on or near the ground. Noted red lettering of some type. Insignia about two and a half feet wide, guess. Was in middle of object like drawing. Object still like aluminum-white.

"Can't tell how long saw object second time (the "close" time). Possibly seconds-just guess-from time got out of car, glanced at object, jumped over edge of hill, then got back to car and radioed as object disappeared.

"As my mike fell as I got out of the car, at scene area, I heard about two or three loud thumps, like someone hammering or shutting a door hard. These thumps were possibly a second or less apart. This was just before the roar. The persons were not seen when I got up to the scene area.

"As soon as saw flames and heard roar, turned away, ran from object but did turn head towards object. Bumped leg on car, back fender area. Car facing southwest.

Officer Zamora next described how he sought shelter as the craft rose up from the ground and then radioed Ned Lopez who was the radio operator back at the police station.

Officer Zamora's report continues:

"As I was calling Ned, I could still see the object. The object seemed to lift up slowly and to get small in the distance very fast. It seemed to just clear Box Canyon or Six Mile Mountain. It disappeared as it went over the mountains. It had no flame whatsoever as it was traveling over the ground and made no smoke or noise."

Officer Zamora completed his statement explaining how he drew a picture of the insignia he observed on the craft which is shown below.

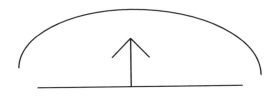

The insignia Officer Zamora saw on the craft is rather interesting. It consisted of an arrow pointed up, topped by a semi-circle and below the arrow was a horizontal line. The entire insignia was red in color and about 2 ½ feet wide.

It's hard to avoid the impression that the semi-circle symbolizes a planet and the arrow represents a spaceship leaving that planet or something along those lines. It would be quite interesting if this same insignia was spotted again on another UFO.

Another implication of that insignia is that its creators have color vision since its color was red. This is perhaps significant in that it is primate type creatures on this planet that have some of the best color vision of all mammals.

Classic Case#: 8

Report Type: CE-I
Location: Exeter, New Hampshire
Witnesses: Teenager, police officers
Date: September 3, 1965

Anyone who has kept track of the UFO phenomenon over the years is undoubtedly familiar with the Exeter, New Hampshire sightings. John Fuller did an excellent job documenting this case in his popular book "Incident at Exeter". I can remember my fascination with this case when I first read Fuller's book since there was such a wealth of solid evidence associated with it.

This case is best described by its three primary witnesses; namely Norman Muscarellowho was a 18 year old teenager at the time and Exeter police officers Eugene F. Bertrand, Jr.and David R. Hunt.

Norman Muscarello's statement to Blue Book was as follows:

"I, Norman J. Muscarello, was hitchhiking on Rt. 150, three miles south of Exeter, New Hampshire, at 0200 hours on the 3rd of September. A group of five bright red lights appeared over a house about a hundred feet from where I was standing. The lights were in a line at about a sixty-degree angle. They were so bright, they lighted up the area. The lights then moved out over a large field and acted at times like a floating leaf. They would go down behind the trees , behind ahouse and then reappear. They always moved in the same sixty-degree angle. Only one light would be on at a time. They were pulsating; one, two, three, four, five, four, three, two, one. They were so bright I could not distinguish a form to the object. I watched these lights for about fifteen minutes and they finally disappeared behind some trees and seemed to go into a field. At one time while I was watching them , they seemed to come so close I jumped into a ditch to keep from being hit. After the lights went into a field , I caught a ride to the Exeter Police Station and reported what I had seen."

The next statement was from Officer Bertrand who was the first patrolman to see the UFO

"I, Eugene F. Bertrand, Jr., was cruising on the morning of the 3rd of September at 0100 on Rt. 108 bypass near Exeter, New Hampshire. I noticed an automobile parked on the side of the road and stopped to investigate. I found a woman in the car who stated she was too upset to drive. She stated that a light had been following her car and had stopped over her car. I stayed with her about fifteen minutes but was unable to see anything. I departed and reported back to Exeter Police Station where I found Norman Muscarello. He related his story of seeing some bright red lights in the field. After talking with him a while, I decided to take him back to where he stated that he had seen the lights. When we had gone about fifty feet, a group of five bright red lights came from behind a group of trees near us. They were extremely bright and flashed on one at a time. The lights started to move around over the field. At one time, they came so close I fell to the groundand started to draw my gun. The lights were so bright I was unable to make out any form. There was no sound or vibration but the farm animals were upset in the area and were making a lot of noise. When the lights started coming near us again, Mr. Muscarello and I ran to the car. I radioed Patrolman David Hunt who arrived in a few minutes. He also observed the lights which were still over the field but not as close as before. The lights moved out across the field at an estimated altitude of one hundred feet, and finally disappeared in the distance at the same altitude. The lights were always in line at about a sixty-degree angle. When the object moved, the lower lights were always forward of the others."

The final statement in Blue Book is from Officer Hunt:

"I, David R. Hunt, at about 0255 on the morning of 3rd of September, received a call from Patrolman Bertrand to report to an area about three miles southwest of Exeter, New Hampshire. Upon arriving at the scene, I observed a group of bright red lights flashing in sequence. They appeared to be about one half mile over a field to the southeast. After observing the lights for a short period of time, they moved off in a southeasterly direction and disappeared in the distance. The lights appeared to remain at the same altitude which I estimate to be about one hundred feet."

In a characteristically poor analysis Project Blue Book attempted to explain this close encounter case as a sighting of some Air Force B-47's conducting training exercises in the area. Of course it is unlikely that B-47 jet bombers would be flying one hundred feet off the ground with their engines shut off, or that the pilots would be able to spend fifteen minutes hovering over a field without power!

Classic Case#: 9
Report Type: CE-II
Location: Nha Trang, Vietnam
Witnesses: American military personnel
Date: June 19, 1966

As a Vietnam veteran myself I found this case particularly intriguing. Nha Trang, the site of this close encounter of the second kind, had a major American military presence during the Vietnam War. I visited the city and its military base several times during my tour of duty in 1967 and 1968 so I am somewhat familiar with the local terrain.

Around 9:45 P.M. on June 19, 1966 the 40,000 men of the Nha Trang military base were treated to an unusual spectacle. High in the sky north of the base a brilliant light suddenly appeared. It was first thought to be an enemy flare until it began to move toward the base at varying speeds. Dropping toward the base it finally stopped, hovering at an altitude of about 300 to 500 feet above the base.

The light was so brilliant it lit up the valley and surrounding mountains as if it were the middle of the day. But the biggest shock came from the object's effect on man made machinery. For about four minutes no car, truck or plane was able to operate at all. The engines on two A-1E Skyraiders, warming up on the flight line, inexplicably stopped. The power all over the base failed and, incredibly, even a Shell oil tanker anchored off shore lost its power. Diesel powered bulldozers cutting roads over nearby mountains also experienced engine failure.

Finally the UFO shot straight up into the sky, disappearing in just 3 to 4 seconds. Full power was restored on the base as soon as the object departed.

It is undoubtedly significant that this sighting occurred during a major UFO wave in 1966.

Classic Case #: 10
Report Type: CE III
Location: Falcon Lake, Ontario/Manitoba
Witnesses: Stephen Michalaq
Date: May 20, 1967

On May 20, 1967 Stephen Michalaq an amateur prospector, was exploring an area near Falcon Lake on the border between Manitoba and Ontario Provinces in Canada. Shortly after noon he noticed two cigar shaped glowing red objects descending toward the ground not far from where he stood. One of the objects landed but the other flew back up into the sky.

The UFO on the ground seemed to cool down, changing from its red color to a silver gold appearance. Michalaq had welders goggles as part of his prospecting equipment and he put these on to protect his eyes so he could inspect the landed craft more closely.

For the next half hour he examined the craft at close range. The craft was about 30 feet in diameter and 12 feet high. The object emitted a bright purple light from its interior and smelled strongly of sulphur. While he was inspecting the exterior of the craft a door opened and Michalaq could hear voices emanating from the aperture. He tried communicating with the voices in several languages but received no response. A little later the aperture closed so Michalaq continued to examine the outside of the craft. He found the outside of the craft to be quite hot, hot enough to melt his glove.

Moving around the object the amateur prospector found himself in front of a feature that resembled some kind of vent. At about this time a blast of hot air shot out of the vent hitting Michalaq in the chest and setting his clothes on fire. Moments later the craft rose up into the sky and disappeared. After extinguishing the burning clothes, he noted that the ground was charred at the spot the UFO had occupied.

Follo wing the encounter Michalaq was hospitalized for first degree burns. The blast of hot air from the vent had left a strange grid pattern of burns on his chest. He also suffered mild radiation poisoning manifested by a blood infection, diarrhea, nausea, stiff joints, a burning sensation and skin infections.   Doctors believed that Michalaq experienced a brief exposure to Xrays or gamma rays but fortunately not enough to be fatal.

I had hesitated to include this case because it has only a single witness who had no special credentials to enhance his credibility. However there were certain features of the case that persuaded me to include it among the classic UFO cases. It is similar to the Air Force lieutenant's case in Nevada where that witness also approached UFOs on the ground. This sighting, in which I have considerable confidence, sets a precedent that suggests it is possible to approach UFOs closely on the ground.   In the lieutenant's encounter he did not get as close to the UFOs as Michalaq did which may have spared him radiation burns.

Another feature that suggested this sighting was authentic was the red color of the object while it was in flight. This coloration is reported often when UFOs are airborne, changing to a different color when the UFOs are on the ground. Finally, in many close encounters witnesses report the sensation of intense heat so it is of interest that Michalaq described the exterior of the craft as being extremely hot.

Classic Case #: 11
Report Type: CE-II
Location: Ohio, U.S.A.
Witnesses: Army helicopter crew and ground observers
Date: October 18, 1973

On the night of October 18, 1973 a U.S. Army Reserve helicopter with a crew of four was heading north from Columbus, Ohio to its base in Cleveland Ohio. The helicopter commander on this practice mission was Captain Lawrence Coyne who had some 19 years of flying experience including numerous night flights. At the controls was co-pilot Lieutenant Arrigo Jezzi. In the back seats were Sergeant John Healey and crew chief Sergeant Robert Yanacsek.

The Bell UH-1 "Huey" was flying at 2,500 feet on a clear night with the ground clearly visible below. Around 11 P.M. Sergeant Healy noticed a red light to the west of the helicopter that was slowly moving south. Several minutes later Sergeant Yanacsek also saw a red light to the southeast of their aircraft that seemed to be pacing them. It was decided this must be a F-100 Super Sabre jet fighter. But within moments the object appeared to change course and head straight for the helicopter! Coyne grabbed the controls from Lieutenant Jezzi and put the Huey into a power dive to avoid what looked like an imminent collision. The commander immediately radioed nearby Mansfield, Ohio and demanded to know if there were any high performance aircraft in the vicinity at their low altitude. However they discovered that both their UHF and VHF channels were jammed and no communication was possible.

The object, however, continued to head directly for them so Coyne took evasive action by increasing the rate of dive to 2,000 feet per minute. Suddenly the object halted in mid air, directly in front and some 500 feet above their aircraft. Looking up the crew could see the object was cigar shaped, about 60 feet long, with a smaller dome on top. There was a bright red light on the leading edge, a bright white light toward the rear and a green light beamed down from beneath the strange craft. The green light beam swung around as if it were a searchlight on pivots and shone directly into their cockpit, illuminating the entire interior of their aircraft with its green light. The green light also reflected off the hull of the UFO.

The object then accelerated rapidly away to the west before changing course to the north-west and disappearing. As it moved further away from them they could only see the white light at its rear.

After the UFO had departed the crew noticed the magnetic compass was spinning wildly. But even more startling was the fact that the altimeter registered 3,500 feet of altitude and indicated they were climbing at 1,000 feet per minute even though the controls were set for a rapid descent! Since earlier in the encounter, during the 27cross27l27 descent, Coyne had seen the altimeter registering 1,700 feet, then somehow the object had pulled their Huey upwards nearly 2,000 feet. When the altimeter registered 1,700 feet that was above mean sea level but the elevation of the Ohio countryside was much closer to the Huey than that. It is interesting to speculate if the crew of the UFO deliberately rescued the helicopter from a dangerous situation.

Five witnesses on the ground confirmed the encounter between the helicopter and the UFO.

After the UFO had left the area, radio communications were restored and worked perfectly. However the magnetic compass no longer functioned properly and had to be replaced.

Classic Case #: 12
Report Type: Radar-Visual
Location: Alaska, U.S.A.
Witnesses: Crew of Boeing 747
Date: November 17, 1986

UFOs have historically shown a particular interest in aircraft and this sighting is not atypical of these aircraft/UFO encounters. For public relations reasons, airlines are generally reluctant to have their pilots reporting UFOs. The pilot of the aircraft involved in this sighting was, in fact, grounded temporarily as a result of his UFO report. This is extremely unfortunate since it discourages the collection of important data that may help us in unraveling the UFO enigma. Fortunately some reports still get through and add to our knowledge of this perplexing phenomenon.

On November 17, 1986 Japan Airlines (JAL) Flight 1628 was traveling from Paris to Japan with stops at various locations including Anchorage, Alaska. The flight was a Boeing 747 cargo plane manned by a three man crew.

Just after 5 P.M. the crew noticed lights off their port side and below the 747. Abruptly the lights shot up to a position in front of the aircraft. This type of rapid maneuvering by UFOs around aircraft has been observed numerous times.

While the lights were in front of the aircraft they appeared like two rectangular displays of bright lights, one above the other. The pilot noted that the cockpit of the aircraft was lit up and he felt a sensation of warmth on his face.

The UFO now moved to the side of the aircraft although two other lights remained in front pacing the 747. The captain turned his on-board weather radar on and it showed a large target about 8 miles away to one side of the cargo plane. On the ground, Air Traffic Control and USAF radars were also showing targets around the 747.

Finally the white lights in front of the freighter fell behind and disappeared from sight and also from the radar screens. The huge Saturn-shaped UFO continued to follow the plane, but as the concerned captain began to change course this giant UFO also disappeared.

Sightings From the 1930's to 1950's
Chapter 3 in Section I

Case #: 1
Report Type: Daylight disc
Location: England
Witness: Building contractor
Date: November 13, 1939

UFOs have been around long before they were called UFOs or flying saucers. This sighting occurred nearly 8 years before the famous sighting by pilot Kenneth Arnold of discs over Mt. Ranier in Washington State that inspired the term "flying saucer". If UFOs are of extraterrestrial origin this is exactly what would be expected. It would be unlikely that UFO technology just happened to parallel the explosion of earth technology during and after World War II when UFOs became very apparent. It is far more likely that UFO technology has been around for a very long time by various civilizations beyond Earth which then began showing more interest in our civilization as our technological achievements became more noticeable.

This early sighting has many of the classic characteristics of later UFO reports which strongly support this as a genuine account.

On the early morning of November 13, 1939 a building contractor was driving by a deserted farm on his way to Brockworth, England. As he neared the farm he heard a high pitched humming sound and moments later observed a bell shaped object which was apparently the source of the noise. The object was dull grey in color with apertures resembling windows part way up from the bottom. Projecting from the bottom was a greenish colored curtain of light.

The contractor stopped his car and got out to watch the strange spectacle. After a few minutes the greenish curtain of light retracted into the object as though it were solid rather than consisting of light rays. As the greenish light retracted into the object he noticed that it appeared to dissipate. At this point the UFO tilted to 80 degrees and moved silently away.

Readers familiar with the better UFO reports will notice various similarities in the descriptions. For example the dull grey color, the window-like apertures in the lower body of the craft, the circular or bell shape, the retracting light beam, etc. These constantly repeated characteristics argue strongly for the reality of UFOs.

Case #: 2
Report Type: Daylight discs
Location: Germany
Witnesses: B-17 bomber crews
Date: October 14, 1943

Small, apparently remote controlled UFOs were seen repeatedly during World War II by pilots and aircrew. Aircrews who had encountered them were convinced they were under intelligent control because of their maneuverings around the bomber formations which clearly suggested an investigation of the aircraft.

Often these devices, called "foo fighters" by Allied pilots, appeared as glowing balls of light. It has been suggested that the pilots had actually seen ball lightening or St. Elmos fire. No doubt this was true some of the time but many pilots were at least familiar with St. Elmos fire and were sure it was not that. And ball lightening is a very short lived phenomenon whereas the foo fighters often maneuvered around the aircraft for extended periods.

If the foo fighters are remote controlled devices operated by the larger UFOs, one would expect them to probably resemble, at least in some respect, the parent UFOs. In fact this appears to be the case.

The larger UFOs are typically shaped like discs, spheres or cylinders. At night and sometimes during the day, the larger UFOs are brilliantly lit, so bright sometimes that it is difficult for observers to make out the underlying structure. On the other hand, many daylight sightings are made of clearly visible metallic looking discs or other shapes.

So if the foo fighters are associated with the larger UFOs then it is not unreasonable to expect a similar technology to be used in both the parent craft and the remote controlled devices. The sighting described next is exactly what would be expected in that the foo fighters were indeed seen to be miniature discs.

On October 14, 1943 the American 348th Bomber Group was on a daylight bombing run over Schweinfurt, Germany. German fighters had left the area, leaving the bombers by themselves. But ahead of the formation of B-17's the aircrews observed a cluster of small discs, perhaps 3 to 4 inches in diameter and one inch thick.

One of the bombers attempted to avoid the discs with evasive maneuvers but was unsuccessful. The wing of the bomber passed through the cluster of discs but there was no damage to the plane.

A disc similar to those reported in the above account was photographed by a test pilot flying next to a B-47 jet in September 1957 near Edwards Air Force Base in California. The year 1957 saw a major UFO wave in the U.S., especially in the fall, so this photograph is undoubtedly genuine.

Case #: 3
Report Type: CE-II
Location: Farm near O'Neill, Nebraska
Witness: Gladys McCage and her 4 year old son
Date: Fall 1946

Mrs. McCage was walking from the barn to her house on a fall evening in 1946 carrying milk buckets with her 4 year old son following her. She noticed a yellowish-orange light in the sky that was heading rapidly toward their location. The object changed to a red color before coming to a halt when it was just about directly over them. Mrs. McCage grabbed her son and fled toward her house.

She described the object as looking like a huge cigar the size of a football field, with windows and making a tremendous noise. But of particular interest she commented that: "….I had a sensation like there was a 'vacuum' and I could have been pulled skyward but I didn't stand still to find that out." She also mentioned a yellowish-blue flame or fire that was shooting out the back of the object and that it had no wings.

As we shall see the appearance of a vacuum effect or the sensation of being sucked up toward the bottom surface of UFOs is often reported in close encounters.

Case #: 4
Report Type: Radar Visual
Location: Near Munich, Germany
Witnesses: Air Force Captain and ground radar operators
Date: November 23, 1948

This incident was the first time a UFO had been picked up simultaneously by radar and observed visually. The report itself went to Project Sign, which was mandated to collect data on the UFO phenomenon as well as conduct investigations.

An Air Force captain had observed an unusual object flying near an Air Force base in Germany. The sighting took place at 10 P.M. with clear skies. The report stated:

"At 2200 hours, local time, 23 November 1948, Capt. _____ saw an object in the air directly east of the base. It was at an unknown altitude. It looked like a reddish star and was moving in a southerly direction across Munich, turning slightly to the southwest then the southeast. The speed could have been between 200 to 600 mph, the actual speed could not be estimated, not knowing the height. Capt. _____ called base operations and they called the radar station. Radar reported they had seen nothing on their scope but would check again. Radar then called operations to report they did have a target at 27,000 feet, some 30 miles south of Munich, traveling at 900 mph. Capt. _____ reported that the object he saw was now in that area. A few minutes later radar called again to say that the target had climbed to 50,000 feet, and was circling 40 miles south of Munich."

The Air Weather Service stated that they had no record of a balloon in this area. And, of course, in 1948 only experimental aircraft would have been capable of climbing 23,000 feet in a few minutes and traveling at 900 mph. However there were no experimental craft in Germany at that time.

Case #: 5
Type: Daylight Disc
Location: White Sands Proving Ground, New Mexico
Witnesses: Launch Crew of a Skyhook balloon
Date: April 24, 1949

On several occasions in the late 1940's, Navy Commander R. B. McLaughlin or his Skyhook balloon launch crew observed UFOs at the White Sands Proving Ground in New Mexico. His crew made a particularly good observation in the spring of 1949. At 10:30 on the morning of April 24, 1949 they had launched a small weather balloon to check lower level winds when one crew member drew their attention to an object in another part of the sky. The man tracking the balloon through the 25 power theodolite turned the instrument on the strange object while the timer reset his stop watch. The object was whitish-silver in color and appeared elliptical in shape. They tracked the UFO for sixty seconds during which time it first descended then rapidly gained altitude. No sound was heard.

From the data collected it was determined that the UFO was about 40 feet wide and 100 feet long and appeared to be relatively flat. Further, they calculated its altitude when it was first observed to be 56 miles high and traveling at over 25,000 miles per hour! While there was some question about the precision of their measurements it is apparent that even if they miscalculated by a factor of 100% the UFO still displayed extremely high speeds and altitudes.

The caliber of the witnesses in this sighting was exceptional since the tracking team consisted of engineers, scientists and technicians. R.B. McLaughlin, the crew commander, after having seen a UFO at White Sands was quoted as saying he was convinced that these discs were spaceships from another planet, "operated by animate, intelligent beings".

Case #: 6
Report Type: Nocturnal Lights
Location: Norwood, Ohio
Witnesses: Scientists, clergy, media, military, etc.
Date: August 19, 1949 to March 10, 1950

Although this UFO sighting is classified as a nocturnal light (NL) it is considerably more interesting than the average NL report.   It is estimated that thousands of people in the area around Norwood, Ohio observed this phenomena that extended intermittently over a period of about 7 months.   Norwood is adjacent to Cincinnati and not far from Dayton, Ohio where Wright-Patterson Air Force Base is located.

The Reverend Gregory Miller of the Saint Peter and Paul Church of Norwood had purchased an U.S. Army surplus 8 million candle power searchlight to be used as part of a carnival celebration held on the church grounds on August 19, 1949.   Sergeant Donald Berger, a member of the University of Cincinnati ROTC was brought in to operate the searchlight.   At 8:05 PM Sergeant Berger caught a large, stationary object in the beam of his searchlight.   The object was described as being solid looking, round in shape, of huge size and appeared to have ridges or ribbing as a structural feature.   Father Miller with the help of Sergeant Berger continued to operate the searchlight for many more months after this first sighting.

During the period they operated the searchlight a number of additional sightings were made, some of which were captured in still photographs and movie film.   Some of the sightings included smaller triangular shaped objects along with the "mother" ship.   Lest the reader be skeptical that UFOs would appear repeatedly in a confined geographic area for many months, it's worth remembering that this has happened at other times such as the Hudson Valley region of New York State in the 1980's and Belgium in the 1990's.

Reverend Miller was able to film the objects on several occasions.   He and others also took a number of still photos.   One of the most important aspects of these sightings was the effect the large mother ship and the smaller triangular objects had on the searchlight beam.   On the night of October 23, 1949 as Sergeant Berger pointed the beam near the mother ship it was observed that the end of the beam bent about 27 degrees (as later measured in the photo plane) toward the craft. The smaller triangular objects also caused the searchlight beam to bend.   This suggests that there are very strong attractive gravitational fields in the vicinity of UFOs.   Still frames from the movie camera film that show the bending of the searchlight beam can be viewed on the internet by doing a search for this 1949 Norwood sighting.

UFOs were a relatively new phenomenon in the late 1940's so at the time the significance of these observations were not truly appreciated by much of the scientific community or the media. But as we shall see these early reports and the unusual effects of UFOs on the environment were part of a pattern that is now apparent in retrospect.

Case #: 7
Report Type: Radar Visual
Location: Korea
Witnesses: Six U.S. Navy Aircrew
Date: September, 1950

This impressive sighting was related to Jim and Coral Lorenzen, founders of the Aerial Phenomena Research Organization, by one of the six Navy witnesses involved. It is a particularly interesting sighting as it includes a great deal of detail and because of the high caliber of the witnesses. Without doubt it is one of the best observations of UFO characteristics that has ever been recorded.

The incident occurred during a carrier launched fighter-bomber daylight mission over Korea in early September of 1950. These Navy fighter-bombers had been assigned the task of strafing and bombing an enemy truck convoy believed to be some 100 miles south of the Yalu River. They never sighted the convoy but what they did see was far more interesting.

The two man crew in each of the three aircraft consisted of a pilot and a radar-gunner. The witness interviewed by the Lorenzens was one of the radar-gunners.

As the witness scrutinized the ground, searching for the target, he suddenly noticed two large, dark shadows racing toward the formation. On looking up he saw the objects making the shadows, they were two huge disc shaped objects. The radar-gunner estimated the objects speed at between 1,000 to 1,200 MPH. At one and a half miles from the formation the radar indicated that the two objects had come to a sudden halt. Using radar and reference points on the canopy it was estimated the objects were 600 to 700 feet in diameter. Visually the objects were observed to "jitter" or "fibrillate" when they stopped in mid air.

The alarmed radar-gunner prepared his radar controlled guns to fire on the objects but as he attempted this the radar was severely jammed. Failing this he tried to radio the carrier but each radio frequency on which he attempted to establish contact was promptly jammed.

The objects, which until now had been pacing the three planes, began to circle the formation, flying above and below the aircraft as if examining them. The six crew members got a good look at the objects from different angles as the strange craft maneuvered around the fighter-bombers. They were shaped somewhat like Chinese coolie hats and had a silvered mirror look with a reddish glow surrounding them. Oblong apertures on the edges alternately emitted copper-green colored light and then a pale pastel-colored light. A shimmering red ring encircled the object above the glowing apertures.

In the center of the bottom of the objects was a inky black circular area that remained motionless as the rest of the object jittered continually. The six crewmen also noticed a sensation of heat inside their aircraft as well as feeling a high-frequency vibration.

After completing their apparent inspection of the aircraft, the two objects departed at high speed in the same northwest direction from which they originated.

Later it was found that the instrument dials on their aircraft had greatly increased luminosity and that all their gun camera film, though not used, was fogged or exposed.

No viable conventional explanation has been found for this remarkable sighting.

Case #: 8
Type: Daylight disc
Location: Artesia, New Mexico
Witnesses: six people
Date: January 16, 1952

Two employees from the Aeronautical Division of General Mills Corporationand four individuals from Artesia, New Mexico, were observing a Skyhook balloon from theArtesia Ariport on June 16, 1952. One of the individuals noticed two small specks on the horizon that seemed to be heading toward the balloon. The two specks turned out to be two round, dull white objects flying in a tight formation. Upon reaching the balloon the two objects circled it and then flew off over the horizon to the northwest. When the objects circled the balloon they had tilted on their side at which time it became clear they were disc shaped.

With the balloon as a reference point it was estimated the discs were about 60 feet in diameter.

Case # 9
Report Type: Daylight disc
Location: Japan
Witness: Air Force pilot
Date: March 29, 1952

The following report is intriguing since it deals with an observation of a very small disc, less than a foot in diameter, similar to case # 2 in this chapter. For years there have been reports of small, lighted objects that seem to be some type of unmanned, remote controlled devices. In World War II they were called "foo balls" or "foo fighters", the word foo is probably derived from the French word "feu", meaning fire.

Since we know that the full sized discs almost always appear as brilliant lights at night and sometimes in the daytime, then it is possible that the small lighted objects are just miniature versions of the larger, classic disc shaped objects. This would imply that the same technology is used for both the remote controlled devices and the manned discs.

This sighting is in the form of an Intelligence Report (IR) submitted to the Air Force's Air Technical Intelligence Center (ATIC) by an Air Force pilot, flying from a USAF airbase in Japan The IR reads as follows:

"At 11:20 hours, March 29, 1952, I was flying a T-6 north of Misawa. GCI (Ground Control Intercept) was running an intercept on me with a flight of two F-84's. One of them overtook me, passing starboard at approximately 100 feet, and ten feet below me. As he pulled up abreast, a flash of reflected sunshine caught my eye. The object which had reflected the sunshine was a small, shiny disc-shaped object which was making a pass on the F-84.

"It flew a pursuit curve and closed rapidly. Just as it would have flown into his fuselage, it decelerated to his air speed, almost instantaneously. In doing so, it flipped up on its edge at an approximate 90-degree bank. It fluttered within two feet of his fuselage perhaps two or three seconds. Then it pulled away around his starboard wing, appearing to flip once as it hit the slipstream behind his wing tip fuel tank.

"Then it passed him, crossed in front, and pulled up abruptly, appearing to accelerate, and shot out of sight in a steep, almost vertical climb. It was about eight inches in diameter, very thin, round, and as shiny as polished chromium. It had no apparent projections and left no exhaust or vapor trails. An unusual flight characteristic was a slow, fluttering motion. It rocked back and forth in 40-degree banks, at about one-second intervals throughout its course."

Note the fluttering motion of the small disc which is a characteristic often reported with the larger discs.

Case # 10
Report Type: Nocturnal lights
Location: Norfolk, Virginia
Witness: Airline pilots
Date: July 14, 1952

On the clear night of July 14, 1952 a Pan American DC-4 was approaching Newport News, Virginia on its way to Miami, Florida. At the controls was First Officer W.B. Nash with Second Officer William H. Fortenberry as co-pilot. The flight was cruising at 8,000 feet.

At 9:12 P.M. the two pilots saw an odd reddish glow ahead of them. Seconds later the reddish glow could be distinguished as six orange-red discs racing toward them at extremely high speed but about a mile below them. Since the discs were only several thousand feet above the ground they were able to accurately estimate their size relative to ground objects. They were about 100 feet in diameter. The discs glowed like red hot metal (a not uncommon description).

The discs were in echelon formation with the leader at the lowest point and the five remaining discs stacked above each other and spread back from the leader.

Appearing to sight the DC-4 the leading disc suddenly slowed, its brilliance dimming as it decelerated. The next two discs in line, seeming not to react as quickly as the leader, wobbled momentarily. Once the formation had stabilized in concert with the leader, all six discs flipped up on edge and changed course by at least 150 degrees. In the new course the six discs quickly resumed the same echelon formation after returning to the flat position relative to the ground.

During the brief moments the discs were on edge it was observed they were about 15 feet thick and dark on the edge and bottom. Only the upper surface was glowing.

Within seconds of this display, two more discs appeared under the DC-4 apparently racing to catch up with the other six discs. As they accelerated to rendezvous with the other discs it was noted they glowed more brightly than any of the discs in the original formation.

At this point all the discs went dark. When the lights reappeared it was observed that the 8 discs were now in line, heading west. After gaining altitude they quickly vanished into the night.

Using a Dalton Mark 7 computer the pilots determined the discs had gone 50 miles in not over 15 seconds. This equated to a speed of around 12,000 miles per hour! Even given some inaccuracies, the performance of the discs was extraordinary.

Case #: 11
Type: Radar Visual
Location: Washington, D.C.
Witnesses: Military and civilian radar operators and pilots
Date: July 19 and July 26, 1952

The Washington UFO sightings created a sensation around the country in the early 1950's. These sightings are not only well documented but were made by highly qualified witnesses backed up by solid radar returns.

This series of sightings, that have since become a classic among UFO reports, began near midnight on July 19, 1952. Two separate radars at Washington National Airport picked up a total of eight targets near Andrews Air Force Base. Commercial pilots in the Andrews area saw strange lights in the same location as the radar targets. The objects behaved erratically, at times moving at only 100 to 130 miles per hour then suddenly accelerating to extremely high speeds. One target was tracked at 7,000 miles per hour. The targets were also picked up by the Andrews AFB radar.

Sometime after midnight a Capital Airlines pilot taking off from National was asked by a controller to keep an eye out for unusual lights. While still in radar range the pilot contacted the controller and exclaimed, "Theres one – off to the right – and there it goes". Simultaneously the controller observed that the unknown, which had been to the right of the Capital Airliner on his scope, disappeared off the screen. In the next fourteen minutes the Capital pilot observed six more identical lights.

Just two hours later another pilot approaching National reported excitedly that a light was following him at his altitude and at his eight o'clock position. Checking their radar the controllers confirmed that there was, indeed, a target behind and to the left of the airliner.

The night of sightings was capped by one final event. In the early morning hours the National Airport tower contacted the Andrews AFB radar station to advise them that an unknown target had appeared on their scope just south of the Andrews tower and directly over the Andrews radio range station. Several Andrews operators ran outside to look and were startled to see a "huge fiery-orange sphere" hovering in the sky directly over the radio range station.

Exactly a week later on July 26, 1952 a similar spate of nighttime sightings occurred in the Washington, D.C. area. This time F94 fighter jets were scrambled to try to intercept the mysterious visitors, some of which were violating restricted airspace in the capital area.

On several occasions jets near Langley AFB in Virginia got radar lock-ons to unknown targets but only to have the contact broken seconds later as the targets swiftly outran the jets. Minutes later unknowns appeared on the Washington National Airport scopes. Again F-94's chased the strange lights but were unable to overtake them.

It is important to note that the July Washington sightings were accompanied by numerous other UFO sightings all along the East Coast of the United States as well as in other areas of the country. For example, F-94's in New Jersey and Massachusetts attempted nighttime intercepts of unknown lights but despite achieving radar lock-ons, they failed to catch the elusive objects.

In Texas a round, silver UFO that spun on its 37cross37l axis was observed racing 37cross the sky. And unidentified amber-red lights were seen over the Guided Missile Long Range Proving Ground at Patrick AFB, Florida.

In the Washington sightings the Air Force had been slow to respond to the numerous sightings, many of which were in restricted areas, and consequently may have been embarrassed by the whole episode. There is some indication that a few Air Force personnel involved in the sightings were pressured to change their original versions of these events. Later explanations that the radar returns were due to unusual weather or temperature inversions are quite unlikely since there were repeated instances of simultaneous radar returns accompanied by visual confirmation of unknown objects. Furthermore the radar operators were highly experienced as were the commercial pilots involved in the sightings. The radar operators were familiar with various weather returns, which are common in the D.C. area, including temperature inversions. Indeed weather returns were picked up on the scopes throughout the sightings but were ignored by the operators.

Case #: 12
Report Type: Radar Visual
Location: Gulf of Mexico
Witnesses: USAF crew of B-29 bomber
Date: December 6, 1952

For those who have followed the UFO phenomenon, the year 1952 stood out as one of the peak "wave" years. This sighting is one of the more interesting ones during 1952 since it clearly reveals the extraordinary performance capabilities of UFOs.

Near dawn on December 6, 1952 an Air Force B-29 heavy bomber piloted by Captain John Harter was returning from a practice mission over the Gulf of Mexico to its base in Texas. The captain asked his radar officer, Lieutenant Sid Coleman, to turn on the B-29's radar so he could examine the Louisiana coastline on his auxiliary scope in the cockpit. They were approximately 100 miles south of the Louisiana coast.

While watching for the coastline on the main radarscope, Lieutenant Coleman was startled to see a unknown target travel 13 miles in one sweep of the radar. Moments later two other targets appeared on the screen. Before the unknown targets disappeared from the scope they were clocked at 5,240 miles per hour!

The captain immediately ordered Lieutenant Coleman to recalibrate his radar scope; no aircraft known was capable of traveling at these speeds in the lower atmosphere.

As Coleman completed recalibration of his set four more unknowns appeared on his scope and the two other scopes aboard the B-29. This time it was clear there were no malfunctions in the radar.

One of the radar unknowns appeared to be swiftly approaching on the right side so the flight engineer, Baily, leapt into the waist blister to see if anything was visible. In the dark night he suddenly saw a bright blue object streak by from front to back, moving so fast it was virtually a blur of light. As before the speeds of the second group of unknowns was found to be in excess of 5,000 miles per hour.

Several more times, small groups of the unknowns raced around in the vicinity of the bomber and once again a blue-white blur was observed by one of the crewmen.

One last group of 5 unknowns appeared on the radar to be racing at breakneck speeds directly toward the bomber from astern. But just as a collision seemed imminent the blips suddenly decelerated to the B-29's speed and for 10 seconds paced the bomber.

Finally the astonished airmen watched the last 5 unknowns on their radar race away at over 5,000 miles per hour and merge, on the move, with a much larger blip on the radar which was also traveling at over 5,000 miles per hour. Incredibly, the large blip then accelerated to over 9,000 miles per hour and disappeared from the scope.

This Gulf of Mexico sighting has many of the common characteristics of UFOs such as high speeds, rapid accelerations and decelerations, brilliantly lighted objects and the pacing of an aircraft.

Case #: 13
Report Type: CE-I
Location: Near Dallas, Texas
Witness: Crews of three B-47 bombers
Date: September 3, 1954

This next report is extremely interesting because of the caliber of the witnesses, the long duration of the sighting and the excellent observations of the UFO. It appeared in the Sept./Oct. 1983 International UFO Reporter whose editor in chief at that time was Dr. J. Allen Hynek

The report was picked up during a radio call-in talk show on UFOs from the widow of a USAF major. Her husband had been involved in a close encounter of the first kind.

The major was a navigator aboard a B-47 bomber engaged in routine bombing exercises over a number of mid-western states on September 3, 1954. The major's squadron of 3 B-47's, with his aircraft in the lead, was returning to its home base, Barksdale A.F.B., in Shreveport, Louisiana While flying at 25,000 feet near Dallas, Texas they received a message from Carswell A.F.B., near Fort Worth, to keep an eye out for reported UFOs in the area.

They then were shocked to see an unknown craft pacing their bomber, just 100 feet overhead. The UFO was a missile shaped object that was a little longer and wider than the fuselage of their own bomber. Oval shaped portholes were visible along both sides, located about a third of the way up from the bottom. It was metallic looking and emitted an orange exhaust flame from the rear. Smaller exhaust ports, which may have been used for directional control, were located a short way forward of the main exhaust. The bottom of the object appeared to be glowing.

The UFO paced the bomber for a short time but then abruptly zoomed ahead and accelerated almost vertically up into the sky until it was lost to sight. Attempting to follow the unknown craft they nearly stalled the B-47.

But suddenly the UFO came back down and began circling the B-47 at a distance of only some 300 feet! It looped above and below the bomber at such high speeds that they could barely see it. It continued pacing them and performing various maneuvers around the bomber for over an hour! At one point the object shot in front of the bomber and executed a perfect figure eight, almost as if it were showing off. Another time it slowed down in front of them but just as a collision seemed inevitable it quickly moved out of the way. Finally the UFO shot up into the sky again but this time it did not return.

Just as UFOs seem to 'toy' with cars they appear also to play with aircraft. The UFO's behavior in this instance hardly suggests serious investigation of the aircraft.

During the long encounter the major and the co-pilot took numerous photographs of the strange object. The major alone shot 32 frames of 35 mm color film with his new Nikon camera equipped with a telephoto lens. Two B-47's flying behind the major's bomber apparently also saw the strange craft.

On landing, the crews of the three B-47's were held at Barksdale A.F.B. for three days of special debriefing on the incident. Unfortunately all the film exposed by the crew was confiscated by investigators and has never been released. This could be an indication of government secrecy around UFO sightings or simply normal security measures. Nevertheless there is a great deal of good observational data in this sighting.

Case #: 14
Report Type: CE-I
Location: New Jersey
Witnesses: Housewife
Date: March 6, 1957

The following report from Blue Book files is an excellent example of a close encounter of the first kind. The witness and her family stated they wanted no publicity in connection with this sighting. The report itself is in the form of a sworn affidavit:

"Even at the risk of being called hysterical, hallucinated or worse, I feel I must make record, on my oath as a woman, an American and a member of the human race, of the following:

"That I saw an airborne object which bore no resemblance to any airplane, helicopter or balloon ever made and flown by man, so far as I know.

"That I was in full possession of my senses and my sanity at the time , and all during the time, of my seeing this object, which lasted for at least one minute.

"That I saw the object at a distance of not more than 150 yards at about 2 P.M. on Wednesday, March 6, 1957, first from a rear window and then from out in the back yard of our home on the road from Great Meadows to Hope, New Jersey.

"That the weather was clear under a low overcast, and the position of the object hovering in the air over the slope below our house was such that I could see it-and hear it-with absolute certainty and with concentrated effort to observe and remember every detail.

"That my attention was first drawn to the object's presence by our dogs barking in their pens behind the house, and by their looking upward at the object, as they and I continued to do so easily until the object departed.

"That the shape of the object closely resembled that of a huge derby hat with a rounded domelike crown 30 to 40 feet high, and at least 50 feet in horizontal diameter above a slightly curled-up "brim" that extended outward for 12 to 15 feet from the bottom of the crown. This brim or bottom surface of the object appeared to be sealed over smoothly and completely in a gentle curve, with no holes or ports or windows or vents of any kind, of which I could see none anywhere else on the entire object.

"That, in the absence of any openings into it, I could see no beings, human or otherwise, inside the object who might be operating and riding in it, and no crew or passengers were visible on the object's exterior during its visit.

"That the color of the object, all over, was a uniform white, dull but clean, with no spots, stripes or other markings whatsoever. Its texture was apparently non-metallic but reminded me strongly of pipe clay.

"That a moderate breeze, from the northeast I think, was blowing in which the object hovered quite stationary except for a gentle rocking motion, like a boat at anchor on water. As the object rocked and in the same cadence it made a low growling or rumbling sound that rose and fell irregularly.

"That beneath the object, extending vertically toward the ground, I seemed to see, and then not see, and then see again, a lot of streamers or lines of some material (or force) that twinkled like the fragile strands of tinsel with which one decorates a Christmas tree.

"That without any marked change of sound except a soft rush of air, sucking away and not blowing toward me, the object abruptly ascended almost vertically, slightly northeasterly at immense speed into the thick cloud cover (maybe 300 feet up [?]) and was instantly gone from sight and hearing.

"Within an hour of my experience my husband telephoned me from New York City. I told him what I had seen, and told him again in fuller detail when I joined him at 7:20 that evening in the city. At his insistence I later that evening repeated my account and answered many questions while my memory was still fresh at the home and in the presence of Mr. and Mrs. _____, 35th Street, N.Y.C. I told my story reluctantly again to several other friends in the next three days (March 7, 8 and 9), but made no formal report to any authorities, fearing ridicule."

The virtual certainty that UFOs are real craft, most likely of extraterrestrial origin, is demonstrated by the obvious similarities in this sighting with other good UFO reports. Certain characteristics are observed over and over again. For example the round shape of the object, the prominent outer rim, the oscillating motion, the tinsel like streamers (in other sightings often referred to as "angel hair"), etc. The reluctance of the witness to report her sighting to the authorities is a fairly common reaction to a UFO sighting and, in my opinion, assures the authenticity of this event.

Case #: 15
Report Type: Radar Visual
Location: Kirtland AFB, New Mexico
Witnesses: CAA Control Tower operators
Date: November 4, 1957

Dr. James E. McDonald of the University of Arizona did a thorough investigation of this sighting sometime after it was investigated by the Condon committee (an Air Force funded study of the UFO phenomenon). This sighting was part of a tremendous wave of UFO sightings seen across the United States in the fall of 1957.

The Condon Committee did a poor job of investigating this sighting, as unfortunately they did on many of the sightings they investigated. For example, in this case, despite the high caliber of the witnesses they were never interviewed by the Condon committee. In contrast Dr. McDonald interviewed the key witnesses in a series of five phone calls as part of a thorough investigation of this case. The following is a summary of his results.

At around 10:45 P.M. on November 4, 1957 the two CAA controllers on duty in the Kirtland AFB tower observed an object descending in a steep dive at the end of one of the runways. The object then proceeded to cross other runways, taxiways and unpaved areas at a 30 degree angle before leveling off and heading toward the CAA tower. As it approached the tower the two controllers estimated the object was only tens of feet above the ground. Viewed through binoculars they were able to see that it had no wings, tail or fuselage. It appeared to be egg shaped, oriented on its vertical axis and had a white light on the bottom.

The object came within 3,000 feet of the tower at which point it stopped completely, hovering above the runway for about one minute. This would clearly rule out an airplane in 1957.

It then started moving slowly again until it reached the edge of the field when it suddenly accelerated upward at an extremely high speed. This would rule out a helicopter.

The UFO was also picked up on the base radar. After leaving the airfield it took up a trail position behind a C-46 that had just left the base and followed the C-46 until both the plane and the UFO moved off the radarscope.

Case #: 16
Report Type: CE-II
Location: Near Proberta, California
Witnesses: Larry Jensen
Date: December 1959

In "The UFO Book – Encyclopedia of the Extraterrestrial" by Jerome Clark there is an interesting close encounter UFO sighting involving a "suction" effect. Larry Jensen was on his way to work early one morning when he noticed interference on his radio and his car lights began to get dimmer. This is often reported in close encounter UFO cases. On getting out of his car to determine the problem he then spotted a very large, brilliant bluish-green crescent like device some distance behind him. The object was around 80 to 90 feet wide, 15 to 20 feet in height and was hovering about 60 feet above the road. During the sighting he suddenly noticed his clothes had become soaking wet. He also had the *sensation of being sucked up toward the object* as if by a magnet. Finally the object departed but Jensen discovered that his battery and generator were seriously damaged.

Sightings From the 1960's to 1970's
Chapter 4 in Section I

Case #: 1
Report Type: CE-I
Location: Dillingham, Alaska
Witnesses: Multiple witnesses
Date: May 19, 1960

A number of Native-Americans observed a metallic, silvery white craft near their Alaskan village of Dillingham in the spring of 1960.   The UFO was shaped a bit like a flattened sphere about 25 feet in diameter and about 10 to 12 feet thick.   Something akin to a flange encircled the midsection of the craft.   Several irregular appendages were also noted.   It came within 200 feet of the witnesses and at times was only about a dozen feet above the ground.

But perhaps the most interesting aspect of the sighting was the physical effect of the UFOon the environment.   A suction noise was heard and at the same time it was noticed that dead grass was lifted high up into the air beneath the UFO.   Two empty 5 gallon cans were also picked up and swirled around under the craft for a distance of about 100 yards.

It seems odd that the UFO would be pulling grass and other debris toward its bottom surface. Instead it would seem more logical that these craft would be creating a force that pushes downward toward the earth's surface.   Nevertheless we must deal with the evidence as it is rather than what we think it ought to be.

Case #: 2
Report Type: Daylight Discs (with mother ship)
Location: Cressy, Launceton, Australia
Witnesses: Minister and his wife
Date: October 4, 1960

In the fall of 1957, during a major worldwide UFO wave, I had observed a cigarette shaped object that was accompanied by a circular object. I noticed that the circular object was no wider than the width of the cigarette shaped object. It was many years later as I became familiar with the literature on UFO sightings that I realized the objects I had seen fell into a particular UFO category. A number of UFO reports described long, cylindrical objects that were associated with gaggles of smaller disc shaped objects. In one report a witness observed the disc shaped objects emerging from the much larger cylindrical object, just as if the discs had been stacked up like plates inside the larger object.

It was immediately clear to me that the large cylindrical objects were something like "mother ships" for the smaller disc shaped objects. Interestingly, the disc shape or flying saucer is the most commonly reported type of UFO. In retrospect I realized that the objects I had seen in 1957were most likely a mother ship with one of its discs. The disc, seen from below, appeared as the round object.

This next sighting report by a minister and his wife fits the pattern of a mother ship with its attendant discs.

On the evening of October 4, 1960 a minister and his wife in the town of Cressy, Australia observed a large, grey cigar-shaped object through a window in their church rectory. Minutes before their sighting some residents of Cressy had reported a mysterious explosion. The reverend estimated that the object was traveling at 500 miles per hour in a straight line course. After flying on this course for approximately one minute the cigar shaped object came to a dead stop about three miles away from the location of the Reverend and his wife. They continued to watch the strange object for another half minute when suddenly they were startled to see five or six small, fast moving disc shaped objects emerge from the clouds near the cigar shaped object. The motion of the disc shaped objects was compared to stones skipping on the water, a not uncommon description of the traditional flying saucer in flight. The discs were estimated to be thirty feet in diameter.

The discs next took up positions within a half mile radius of the mother ship. Moments later the entire formation departed in the direction from which the mother ship had originated, disappearing in rain clouds in less than half a minute.

Case #: 3
Report Type: CE-II
Location: Kansas
Witnesses: Two drivers
Date: August 4, 1965

Around 1:30 AM on the night of August 4, 1965 a trucker was heading toward Abilene, Kansas when a UFO flew over the top of his truck and landed on the highway in front of him forcing him to brake hard. A car coming from the opposite direction had to turn sharply to avoid hitting the landed UFO. The truck's diesel engine continued to run but the headlights failed. At this point the UFO moved further down the highway and as it did so the truck's lights came back on. The truck driver was now able to see the UFO and described it as looking like a domed disc about 15 feet in diameter that was hovering a few feet off the ground. Just as the other driver came over to speak to the truck driver the UFO emitted blue sparks from its underside and departed the area.

Case #: 4
Report Type: CE-II
Location: Wanaque Reservoir, Wanaque, New Jersey
Witnesses: Police officers, town officials and others
Date: Throughout 1966

There were a series of extraordinary UFO sightings around the Wanaque Reservoir area of northern New Jersey in 1966 that I believe revealed vital clues to the physics behind the UFO phenomenon.   The caliber of the witnesses and the long duration of some of the sightings makes these reports highly credible.   These sightings were documented in considerable detail in a book written by the editors of Science & Mechanics magazine (now defunct) in 1968 titled "The Official Guide to UFOs."

The sightings occurred throughout 1966 but the most interesting observations were made in October of that year. Sergeant Ben Thompson of the Wanaque Reservoir Police was interviewed by editor Lloyd Mallon about the UFO sighting he had witnessed sometime after 9 P.M. on October 11, 1966. A brilliant disc shaped object with a dome on top maneuvered silently above the reservoir for about three minutes. Thompson estimated that at the beginning of the sighting the object was about 250 feet above the reservoir and only 250 feet away from his position on the shoreline. Part of his interview with editor Mallon is included here:

Sergeant Thompson: "And it went over the trees – which would be on the mountain to the west – it would sort of pull the tops of the trees together. In other words, it had a suction effect. It didn't blow the trees apart. It pulled them together. And it also pulled the water upward."

Editor Mallon: "This sounded incredible to me. 'Can you describe that effect in more detail?' I asked Sergeant Thompson."

Sergeant Thompson: "And as this thing faded away, from an area like, say, to the west of one mountain, as it went over the reservoir to the east, I could see the water come up toward this flying object. Then as the object moved away from that area, the water would settle back down to its natural level."

Editor Mallon: "You mean that the water would move like a wave, or a quick tide?"

Sergeant Thompson: "No. The water was pulled up. It was sucked upward. But not off its bed. The flying object would just raise a whole big area of water – I don't know – for maybe two-hundred and fifty feet. As far as I could see. The object would just pull at the water and I could plainly see the water rising. And when this thing flew away from the area, the water would just settle right down again."

Sergeant Thompson continued: "And that object just pulled the trees right together. The tops of the trees came *right* together. Each tree just mingled in with the other one. They came together just as smoothly as could be. It wasn't a violent motion. It didn't break the trees or anything like that. It would be just like somebody took a big rope and circled around four or five hundred trees and then ran it through a chain-block and started pulling those trees together. And they'd come together nice and slow. Well, that's the way those trees acted when the flying object passed over them."

Editor Mallon: "How high would you estimate that it pulled up the water?"

Sergeant Thompson: "Oh, I would say that from where I was standing – while I was looking into this light – that it pulled the water up a good two or three feet. The reservoir was low at the time and I could see the water rise plainly."

A woman driving her car nearby also witnessed this same UFO but was so frightened that she turned around and returned home.

The truly curious thing about this sighting is the suction effect or attractive gravitational force that pulled the trees and water *toward* the object.   Nevertheless this UFO report provides crucial information about UFO technology that will help us solve this enigma.

Case #: 5
Report Type: CE-II
Location: Burkes Flat, Victoria, Australia
Witnesses: Ron Sullivan
Date: April 4, 1966

On the evening of April 4, 1966 Ron Sullivan was driving his car along a straight section of road near Burkes Flat in the state of Victoria, Australia.   He was startled to see several brilliant white oval shaped objects between 15 and 20 feet in diameter, one above the other in a field just to the right of the road.   Sullivan was then surprised to see his headlight beams bending toward the intense light of the object.   As he got closer to the object the angle of bending got even more acute causing him to turn his car in the opposite direction since he thought he was driving off the road.   Instead he came close to hitting a tree on the left side of the road.

This case is of great interest since it gives the impression that UFOs are surrounded by a strong attractive gravitational field.   This case is similar to the Norwood, Ohio sighting in 1949 where a searchlight beam was seen to bend toward a UFO hovering thousands of feet above the town.

Case #: 6
Report Type: CE-I
Location: Beverly, Massachusetts
Witnesses: Multiple
Date: Spring, 1966

It is difficult to explain this case in any way other than an alien craft under intelligent control. The multiple witnesses included a number of adults as well as two police officers. This case also demonstrates the rather frivolous behavior of these extraterrestrial visitors which is noticeable in many other close encounter cases.

On a spring evening of 1966 a number of lights were spotted over a high school in Beverly, Massachusetts. Three women who had been observing these lights noticed a disc shaped object descending in their direction. The disc was the size of a large car, dull aluminum in color and had a bright light on top. The UFO descended until it was directly over the women at which point two of the frightened women ran out from under the craft. The third woman, immobile from fear, put her hands over her head thinking the object was going to crush her. The UFO, only 25 feet above this woman, then tilted slightly and moved off to a position directly above the high school.

Several of the neighbors as well as two police officers arriving on the scene also observed the UFO. The police report described the object as "like a large plate hovering over the school". The police report also noted that the object had three lights; red, green and white and that it made no noise.

The entire sighting lasted 45 minutes which makes it virtually impossible that the witnesses had observed a helicopter or other type of aircraft.

Case #: 7
Report Type: Daylight Disc
Location: Alberta, Canada
Witnesses: Warren Smith and 2 others
Date: July 3, 1967

On a late afternoon in the summer of 1967, while returning from a hunting trip, Warren Smith and his two companions observed an object descending toward the ground. Though no noise was heard they thought it was a plane in trouble with its engines apparently shut off. Smith quickly retrieved a loaded camera he had in his backpack and began snapping pictures of the falling object. At about this time it was realized that the craft was not an airplane since it had no wings, rather it seemed to resemble a metallic disc. The disc dropped behind some trees but then moments later it reappeared again climbing into the sky. It then hovered briefly, appearing to drop a small object, before rapidly disappearing in a southerly direction.

This incident was thoroughly investigated by Dr. J. Allen Hynek who concluded that an unknown object had been observed. Hynek felt that a hoax was unlikely in this remote, forested area southwest of the city of Calgary.

The photographs were analyzed by the Defense Photographic Interpretation Center of the Royal Canadian Air Force who concluded that the disc was some 40 to 50 feet in diameter and 11 to 14 feet thick. The disc image exhibited a "softening" effect that occurs when something is seen at some distance through the atmosphere. This indicates the object was quite large based on the size of the photographic image.

This sighting contains the classic description of a disc that is so often reported but it also includes excellent photographic evidence.

Case #: 8
Report Type: Daylight Discs
Location: Cambridge, Massachusetts
Witnesses: Two employees of research firm
Date: Fall, 1967

The years 1966 and 1967 saw a major wave of UFO sightings in the United States and around the globe. This sighting occurred during the 1967 wave. In talking to people randomly I am always struck by the fact that when they have a UFO story to relate, that the sighting has more often than not occurred during a major wave. The witness or witnesses are usually not aware of this connection but the fact that the sighting has occurred during a major wave adds additional credibility to their report.

This next sighting is from Carolyn Redmond who was my sister's roommate in Cambridge, Massachusetts in the 1960's. Carolyn is a scientist so she makes an excellent witness based on her analytical background.   Here is the letter she wrote to me regarding her sighting:

"September 25, 1995

Dear Bob,

Re: UFOs sighted in Harvard Square.

It was a clear, sunny, fall day in 1967 about 1:20 P.M. when I looked over toward Harvard Square. Suddenly, some silver, oval discs in the sky above my building caught my eye. As I looked up I saw about 15 discs fly overhead, about 150 feet above the ground. Looking up, I could not see what was flying over the building, where my vision was obscured by the building. I was so startled by these discs that I said to the secretary in the room, 'Look, Charlotte, UFOs.' She jumped up and raced the 8 feet to the window and was in time to see at least 15 of the discs herself. They were flying about 40 MPH with a distance of 8-10 feet between them. They seemed to be about 10 feet in length and about 8 feet wide with a small, round under belly. We were looking up at the underside of the discs. (The window was narrow so two people could not look out at the same time.) Since we had each seen about 15 discs each, both looking out the window at separate times, the total of what we saw was at least 30. The episode took less than 30 seconds and they were out of our view. We both looked at each other, absolutely startled, amazed, and wide eyed, not knowing what to make of what we had seen. We both agreed that there were NO WINGS on the discs. Our reaction was that 'it must be military' since we are quite close to Hanscom Air Force Base. But it was not like anything we had ever seen. It was a clear day and I could see two planes in the sky, just taking off from Logan Airport. American Airlines and Air Canada were clearly visible on the sides of the planes. The sanitary inspector in my building, called the airport to ask if any unusual sightings were reported by those two pilots. Of course at that time no pilot would confess to seeing UFOs - they would lose their jobs. I was not aware of any noise from the discs, but then we were in a room with no way to open windows, or hear much outside noise. I do not have any answer for what we saw and I have never seen anything like it again.
Sincerely,

Carolyn Redmond"

    Carolyn also added the following postscripts:

    "Since I had been trained as a Microbiologist and was working as a Bacteriologist at the time it was necessary for me to be exact and precise when working with specimens. Had no one else seen what I had, I might have wondered if they were real - but I had confirmation by someone else in the same room. My eyesight was excellent since I could see the writing on the airplanes at quite a distance away, as they headed west.
    The discs were flying from west to east - along Mt. Auburn St. but I don't know where they went after I ceased to see them. My window was facing north."

Case #: 9
Report Type: CE-II
Location: Cochrane, Wisconsin
Witnesses: Two women in a car and others.
Date: April 3, 1968

This UFO sighting is of particular interest because of the physical effects the two primary witnesses experienced. These physical effects are important because they give us insight into the probable technology of these craft.

The sighting occurred one evening near a small town in Wisconsin in the early spring of 1968. The witnesses were a schoolteacher and a former US Air Force stewardess who were traveling in a car on an isolated country road. Dr. J. Allen Hynek investigated this case and the schoolteacher (the driver) provided the following testimony as quoted in Dr. Hynek's 1972 book "The UFO Experience – a Scientific Enquiry.": "…..that thing came from the dip in the hill, real fast but real, real smooth like something gliding, but lower than any plane, and hovered and stopped above that car (a car that had just previously passed the observer's car). Then is when it's (the other car's) lights went out, and I pulled onto the gravel because I thought it was a kid. He put out his lights, and I didn't want to smash into him – at all of this my lights were dimming slightly, but I didn't think anything of it until my engine, lights and radio went out and stopped. This happened to me when it (the UFO) left that car and came down the highway….and was above us. It came down over from the other car. It was pretty low. When I looked out my windshield I had to bend forward toward the wheel, and I looked straight up and there it was above us – with the car dead. I had opened the window when the other car's lights went out, and it was open then – and absolutely no sound."

The interview with the witness continued and she described the object as crescent shaped with a more solid appearance in the center part but fuzzy toward the edges. The color of the object was an orangish-red.

But the most interesting testimony was the physical effects experienced by the witnesses. Dr. Hynek commented on these effects: "(they are) reported in other cases also, which may point to the physics of the UFO." As we shall see later in the book I believe Dr. Hynek is exactly correct. Following is an excerpt from the testimony of the schoolteacher on the physical effects that she and her passenger experienced:

"…you know, if you stay in a house at night and everything is still, there are still the noises of the living, you know, but when this thing was there, there wasn't even the noise of the living. It was nothing. It was an eerie quiet. …Another thing I remember…as though I was light in weight and airy. Something like the first time you experience an airplane takeoff or drop from an air pocket. It felt like the air and everything was light and weightless."

This sense of weightlessness has been reported by witnesses in other cases where they have been directly under a UFO. This is truly puzzling since you would expect that the UFO would be exerting a downward force on objects beneath the craft – a kind of anti-gravity if you will. Instead witnesses report a sense of weightlessness or being pulled up toward the UFO. But as we shall see this physical effect is not outside the direction modern physics is taking in its quest to explain nature at its deepest levels.

The eerie silence and stillness is also frequently reported in close encounters of this type. I believe this witness testimony is providing vital clues to UFO technology.

Case #: 10
Report Type: Radar Visual
Location: Airliner enroute from Phoenix to Washington, D.C.
Witnesses: Flight crews and FAA traffic controller
Date: June 5, 1969

This report is about as good as you can get as far as the qualifications of the witnesses are concerned. In this case a FAA flight controller had joined the crew of an American Airlines Boeing 707 in the cockpit for a familiarization flight related to FAA procedures. The sighting was reported to NICAP (National Investigations Committee on Aerial Phenomenon) by the FAA controller.

On the afternoon of June 5, 1969 the American Airlines jet was on a scheduled flight from Phoenix to Washington. D.C. The flight was at 38,000 feet flying east in the vicinity of St. Louis, Missouri with two other aircraft on the same course following not far behind when four unidentified objects suddenly appeared ahead of the American flight. One object was larger than the others, appearing to be about eighteen to twenty feet long, eight feet thick in the middle and perhaps twelve to fourteen feet across the rear end. The three other objects were smaller and pointed in front, somewhat resembling darts. All the objects had the color of burnished aluminum. The St. Louis FAA Control Center advised the American flight that they had picked up the unknowns on their radar.

The objects were in a square formation and traveling on a flight path that put one of the objects on a collision course with the 707. The crew were then startled to see the three smaller objects quickly move closer to the larger object after which the entire formation passed the 707 at a safe distance. The two aircraft following the American flight also had similar close encounters with the unidentified objects. In each case the objects flew head on toward the other aircraft and only at the last second avoided a collision.

Case #: 11
Report Type: CE-II
Location: Lemon Grove, California
Witnesses: Two 11 year old boys
Date: November 16, 1973

The following 1973 close encounter was investigated by Donald R. Carr, a Mutual UFO Network (MUFON) state section director for California. It is an unusually interesting case since it gives possible clues to the technology that UFOs utilize. I italicized part of the witness's statement. Carr filed this case report with MUFON:

"The witnesses were two boys, both age 11. While playing outside, the boys went down into a vacant area next to the group of four houses in which they live. This area is about 80x100 feet, is surrounded by a chain-link fence, contains a couple of small trees, and the ground is composed of hard clay covered by dead field grass. The area is surrounded by several neighboring houses, which are about 150 feet away.

The boys passed through private property on the way to the field. After passing a clump of bamboo, they came out into the open, and saw an object sitting in the darkened field. They slowly approached it and, after about five minutes, one boy who had a flashlight in his hand walked up to it and rapped on it three or four times with the flashlight. The object, immediately came to life. The rapping had produced a metallic sound.

A dome on top of the object, about as high as its diameter, became illuminated with intense red light which irradiated the entire area, including the boys. At the same time, the object, which had been about 18 inches off the ground, rose up to about three or four feet. A row of green lights around the peripheral rim of the object started to blink in sequence and the object started to rotate making a not very high pitched sound which sounded like 'woooo shooo woooo shoo.' The rate of rotation became very high with the red light blinking on and off. Then the red light went off momentarily, came back on, and the object rose into the air, still making the same sound.

The boys started to run, felt chills, tingly and weak. They agreed that they felt as if they were going to black out and *that they were running in slow motion*. They said the object took off toward the Southwest and, after they got up to the street, they saw it disappear into the clouds."

The size of the object was later estimated to be about 20 feet wide with a height of about 10 or 11 feet. The boy's story was corroborated by the discovery of three depressions in the ground that formed an equilateral triangle measuring about six feet eight inches between centers. The indentations in the hard clay soil were up to six inches deep suggesting a craft of great weight. Houses in the neighborhood also reported severe TV interference at the estimated time of the boys encounter with the UFO.

But the most interesting aspect of the UFO report is the observations the boys made while at close range to the object. When the craft switched on its "motor" the boys started running away and as they fled they experienced the sensation of being in "slow motion." This suggests a strong gravitational field in the vicinity of the UFO.

The boys also noticed a row of green lights blinking in sequence around the rim of the craft. As other UFO researchers have noted the outer rim portion of the UFO seems to be where it "does its business."

Case #: 12
Report Type: Radar Visual
Location: Tehran, Iran
Witnesses: Iranian Air Force pilots and ground observers
Date: September 19, 1976

Although Project Blue Book is now defunct, the Department of Defense apparently still maintains an interest in UFOs. This interesting report from Iran was included in files retrieved under the Freedom of Information Act from the Defense Intelligence Agency (DIA). While some think this is an indication of a high level conspiracy in which the US Government operates some sort of secret organization for studying UFOs, it could also be just more or less random interest in the UFO phenomenon by various government agencies. However the DIA documents on this sighting does give it far more credibility than say merely a newspaper account.

The series of events that constitute this sighting started around 12:30 AM on the night of September 19, 1976. At that time the Iranian Air Force began receiving reports of a bright light hovering in the sky near Tehran that resembled a gigantic searchlight. The duty officer took a look outside and saw the strange light himself and quickly requested a fighter intercept.

An F-4 Phantom was launched at 1:30 AM heading north toward the unknown light. Although the light was estimated to be 70 miles away it was so bright the pilot had no problem locating it. However the F-4 pilot was unable to catch up to the object and eventually had to call off the chase. But as he returned to base he was shocked to see something following him at terrific speed, within moments catching up and passing close by the F-4. All instrumentation, including radio communications, were out of commission during the encounter.

A second F-4 had been launched about 10 minutes after the first F-4. This Phantom managed to get a radar lock-on and began to gradually close on the unknown target. But shortly after lock-on was achieved the target suddenly accelerated to a velocity beyond the maximum speed of the F-4.

While still chasing the object the two man crew on the second F-4 were startled to see a ball of light, about 12 feet in diameter, ejected from the object they were following and coming directly toward their aircraft. The pilot reacted instantly by triggering a Sidewinder infra-red missile. But nothing happened, like the other F-4 all instrumentation and communication failed. Now the pilot had to take rapid evasive action so he maneuvered the aircraft into a fast negative G dive but the ball of light continued to pursue the Phantom. Incredibly, the ball of light performed a U-turn inside the emergency evasive turn of the F-4 and returned to the parent object. All power was now restored to the fighter but the pilot prudently decided not to attempt to fire any more missiles at the objects.

As they watched the original object they were amazed to see a second ball of light ejected by the strange craft which headed straight for the ground below. This ball of light seemed to land then spread light over a huge area. While this was going on the parent object accelerated to a much higher speed and disappeared within a few seconds. Meanwhile the light on the ground faded, leaving the whole area in darkness.

But this rather bizarre encounter was not quite over yet. On returning to base the pilot had to circle the base a number of times to regain his night vision due to the brilliance of the unidentified lights. At a certain point in these circuits he lost radio communication each time. A civilian airliner reported the same effect.

Later investigation revealed that a family living in the isolated area where the second ball of light had touched down reporting hearing a loud noise and had observed a brilliant light at the time of ther fighter/UFO encounter.

Phillip Klass, a UFO skeptic, has suggested that the electronics on the second jet failed due to faulty maintenance and that the pilots were merely chasing stars or meteors. But it seems unlikely that both F-4's and a civilian airliner would all have similar electronic and radio problems on the same night. And it is even more improbable that the pilot of the second jet would suffer the loss of his night vision from stars or meteors.

The balls of light emitted by the parent UFO have been observed in other UFO sightings and seem to be used for defensive or offensive purposes.

More Evidence: The 1980's into the 21st Century
Chapter 5 in Section I

Case #: 1
Report Type: Nocturnal lights
Location: Billerica, Massachusetts
Witness: High School Science Teacher
Date: Autumn, 1981

The most commonly reported UFOs are disc shaped objects, spheres, cylinders or craft resembling flying wings. Generally these UFOs seem to be of sufficient dimensions to carry human sized occupants. But there are also a sizeable number of reports that describe much smaller objects that are too diminutive for a human sized crew. Typically the smaller objects are estimated to range in size from a few inches in diameter to several feet.   These miniature UFOs have many of the characteristics of their larger brethren and the obvious assumption is that they are some kind of remote controlled devices using the same technology as the larger "manned" craft.

The following report is from a good friend, Mark Petricone, whom I have known for over twenty years. Mark is a high school science teacher as well as a dedicated amateur astronomer. Like many well qualified witnesses he was baffled by what he saw and discouraged by the apparent indifference of authorities to his unusual observation.

What Mark did not know is that his sighting was part of the larger pattern described above. In World War II , for example, these smaller objects dogged aircraft from all the warring nations and were referred to as "foo balls" or "foo fighters" by Western flyers. Interestingly, each side thought the devices were secret weapons operated by their opponents!

The reaction of various authorities to Mark's sighting is fairly typical since the topic of UFOs does not yet have official respectability in much of the scientific community or among our cultural leaders in general.

Following is Mark's report in his own words:

"It was a Sunday night in Autumn in 1981. I live in Billerica, Mass. within 5 miles northeast of Hanscom Field/Bedford airport. I had just arrived home and was now taking my dog out for her nightly walk. The time was around eight or nine o'clock. I put her leash on, went down the stairs and started up the driveway up to the street. As I reached the street, I glanced to the southern sky at a jet in the usual landing pattern moving west to east. After moving to this house in 1981, I have seen countless planes flying in the same part of the sky moving in the same direction, west to east. I still watch this routine to this day (over 14 years later). Sometimes it's like a parade of aircraft so as one disappears in the trees to the east another appears from the trees in the west. I have a relatively clear view to the south.

"Being a sky watcher since I was 5 years old, I never resist watching whatever happens to be in the sky above or around me. I have always had an interest in astronomy and being a high school science teacher, I even teach a general Astronomy course to juniors and seniors. I own my own telescope and many times I have braved the cold to photograph or just observe something of particular interest in the night sky. I have observed satellites and orbiting spacecraft, meteors, a few comets, planets, eclipses, the northern lights, rainbows, moon halos, sun halos, sun-dogs, sunspots and many deep sky objects such as nebulae, star clusters and galaxies. I can recognize most of the constellations and explain the motion of the moon and stars through the seasons.

"On this particular night I saw something that I just could not explain. As I walked my dog to the top of the driveway, my eyes were drawn to another plane on the usual route in my southern sky. It traveled west to east. However there was something odd in the scene. Traveling right behind the plane were seven reddish-orange dots. They were strung out a good distance so that the nearest one to the plane may have been about a plane-length behind and then the rest were interspersed between this one and the last one that may have been six to eight plane lengths behind. My first reaction was that the plane was letting out sparks since all seven reddish dots seem to follow right in the path of the jet. There was no symmetry to the pattern. A couple were closer together and one or two were a bit lower than the rest. However they all seem to follow the aircraft. I stopped the dog and kept watching silently simply because I had never seen this before. As the plane moved more to the east, it was starting to move into the thin branches of the treetops. I could still watch because the leaves had fallen off the trees. It was then that two of the orange dots moved left and away from the aircraft as the rest stayed on their path behind the plane. I moved to my right dragging my dog with me as I strained to watch this sight another second or two. Soon the branches thickened and I could see nothing.

"At this point I realized I saw something very strange. Without letting my dog do her business I walked right back into the house and picked up the phone. One must understand that I have always hesitated to use a phone. I always seemed to get tongue tied and embarrassed when I couldn't find the right words to speak my thought. I always asked my wife to call even if it was to order a pizza or Chinese food for take out. My wife was upstairs doing something I cannot recall. I dialed 411 and asked for the number of Hanscom field. Without a second of hesitation I was dialing for an explanation. The voice at the other end said 'Hanscom Field' or something like that. I said 'yes, ... I was wondering if you could help me. I just saw something in the sky in the direction of Hanscom field and I was hoping that you could tell me what I just saw'. His response was 'could you hold please?' Without any choice I was holding the phone and waiting for the person to come on to talk to me so I could at least tell him what I saw. In a minute or two the voice was back and without letting me speak he asked 'red lights?' I responded with 'yes, I just saw a plane go by that was followed by seven red lights'. He broke in again, 'yes, someone from Waltham just called about the same thing......I called up the tower but they didn't see anything'. Then I said 'well listen, I have never seen anything like this before... who could help explain this to me?' He said 'well, nobody here saw anything, you may want to call Logan Airport to see if they got anything on the screen'. I said O.K. and hung up the phone. I called information again and before I knew it was dialing Logan Airport in Boston. This person also said that 'nothing unusual had been seen tonight' and really gave me the feeling that he didn't want to spend too much time talking to me. I repeated my story about the seven lights and I insisted that I saw something very odd. I explained that I was a long time sky watcher and this was no crank call. He suggested that I call my local police Department and that's when I got irritated. I repeated 'listen I am not a nut, I just saw seven red lights following an aircraft moving west to east in the Bedford area, I just want to know what I may have seen? ..... the Police department are not going to know what I saw'. He said 'you could also call Pease Air Force Base in Southern New Hampshire'. I hung up and just for the heck of it called the local police. They had received no calls about strange lights. I told them how I had called Hanscom and Logan and they suggested I call the police. The desk officer just humored me and politely repeated 'no calls like that tonight'. By then I was feeling frustrated. After explaining the whole story to my wife I gave up for the night. I went and walked the dog and kept gazing to the area of sky that had had the strange lights.

The next morning I went to school, but I could not stop thinking about what I had seen. During my lunch break I got to a phone and called the Harvard Observatory - Center for Astrophysics. A nice lady answered my call and I repeated the whole story to her. She politely told me that they do not do any collecting of UFO reports and offered the telephone number of a UFO club that I may wish to contact. I was again flabbergasted. I began again...'I am a science teacher...I am an avid sky watcher...I know some astronomy...I am not a nut...I saw something very strange last night and I just want someone to tell me what it may have been!!!!!!' She was polite but she simply said 'I'm sorry, I do not know what to tell you to do'. I thanked her for her time and hung up. I made one last call to Pease Air Force Base in New Hampshire. They told me that nothing unusual had been reported the night before but if I wanted to call and talk to the night crew, to call that night......I never bothered. I finally understood how people who have reported strange lights etc. must be made to feel. Everyone just thinks that you made it up or it was something very simple that you must be too stupid to understand. I do not believe that aliens are visiting but I still wish I knew what it was that I saw that night. I guess someone in Waltham and myself have the same wish."

Case #: 2
Report Type: Nocturnal light
Location: Charlottesville, Virginia
Witnesses: Two adults
Date: March 30, 1982 and April 1, 1982

This case appeared in the November/December 1983 issue of the International UFO Reporter published by the Center for UFO Studies and was investigated by Dr. J. Allen Hynek. It's included in our selected UFO cases since it illustrates the sometimes odd, seemingly playful behavior of UFOs with automobiles.

The principal witness was a young woman living with her parents in Charlottesville, Virginia. She worked as an administrative coordinator for the Neuroscience Program at the University of Virginia Medical School. The second witness was the mother of the principal witness.

The young woman was driving home one night when she noticed a large and brilliant light in the sky ahead of her as she neared the subdivision where her parents lived. The light was estimated to be about one quarter mile away and several hundred feet high.

As she continued down the road she realized the object was stationary, hovering above the road. Thinking the intense light might be from an airplane she rolled down the window of her car but could hear no sound.

Turning left at an intersection the young woman was startled to see that the light had passed over the roof of her car and taken a position on the driver's side of the car where it was easily visible to her. The brilliant light continued to pace her car for the next 10 minutes of her ride home, making every turn she made and staying in the same relative position. No other traffic passed her during this period.

Although the object was too bright to determine its exact shape she was able to distinguish at least five separate sections which were each emitting brilliant white light.

When the young woman stopped in front of her house, the brilliant object also stopped, hovering above the house across the street and now only 400 feet away. Hastily retreating into her house she was able to bring her mother to the front of the house to witness the object gently gliding away over some tall pine trees.

Two nights later there was a sequel to the first sighting. The principal witness was on the phone with Dr. Hynek from 10 to 11 P.M. discussing the sighting, after which she settled down to watch the TV sitcom *MASH*. Glancing out the window around 11:30 P.M. she was shocked to see the huge, brilliant light again! Just as her mother arrived at the window a few moments later, the object flared up in brightness and sped away into the dark night. This sudden increase in luminosity is often reported in UFO sightings as the UFO goes from a stationary position to rapid acceleration.

Although mother and daughter rushed outside immediately after the object flared up, they were unable to see the object again. There were no more sightings after the second incident. The return of UFOs to the scene of earlier sightings is, however, one of the ongoing puzzles in the UFO enigma.

Case #: 3
Type: Nocturnal lights
Witness: Jim Cooke
Location: Croton Falls Reservoir, New York
Date: October 28, 1983

Jim Cooke in the fall of 1983 worked as a biomedical engineer in the field of laser optics and laser surgery. Cooke was driving along Croton Falls Road in Croton Falls, New York on his way home to Mahopac, New York, when he noticed lights in the trees. The lights were heading toward Croton Falls Reservoir, ahead on his left, and seemed to be descending much too fast to be an aircraft. Cooke stopped his car by the reservoir and got out to take a closer look at the strange lights. Emerging on the lake shoreline from the woods surrounding the lake, he was startled to see a large triangular shaped object about 200 feet away hovering some 15 feet above the water. The silent object appeared to be 100 feet in length but Cooke could not make out any details on its surface.

At this point 9 red lights lit up along the sides of the object, and then a glowing red probe or light projected from the bottom of the object into the water. The object moved to several other locations in the reservoir and each time the red probe would extend to the water below.

Cooke watched for about ten to fifteen minutes. Several times during this period cars passed by the reservoir and each time that this occurred all the lights on the object blinked out.

Finally the object soundlessly rose up into the sky and disappeared.

Jim Cooke's sighting was only one of several thousand observations of triangular shaped objects seen in this Hudson River Valley area just north of New York City during the early to mid 1980's. Sightings continued after that but at a slower pace.

Case #: 4
Report Type: Daylight Disc

Location: Mediterranean Sea, 20 nautical miles from Gibralter
Witness: Crewmen on a Russian tanker
Date: June, 1984

On a late afternoon in mid-June, 1984, two crewmen on watch aboard a Russian tanker near Gibralter spotted an object flying toward the stern of their ship. The object looked like an upside down frying pan and emitted brilliant flashes of light. When first sighted it was about 2 nautical miles from the tanker and under 5,000 feet in altitude.

The object rapidly approached the ship and then once abreast of the tanker, it slowed down and paced the ship for 3 minutes. After this the object slowly moved away until it apparently spotted another ship coming toward the Russian tanker. The unidentified object quickly flew over to the other ship, descended to about 1,500 feet above it and hovered for about a minute and a half. Most of the Russian ship's crew was watching the aerial show by this time.

The captain and chief mate of the Russian tanker radiod the other ship, identified as an Egyptian cargo vessel, and confirmed that the Egyptian crew had also observed the strange object.

Finally the object headed back toward the Russian ship before ascending rapidly and disappearing in the clouds. The entire sighting had lasted 12 minutes.

During the sighting the two crewmen on watch got an excellent look at the object through their binoculars. They described it as circular or disc shaped and about 66 to 82 feet in diameter. It appeared to be made of 2 sections that rotated opposite one another. Around the edge of the disc there appeared to be lighted portholes.

When the UFO had first approached the Russian tanker, it had been gyrating which gave the two men on watch a good view of the upper surface of the craft. In the top center portion of the craft was a large red cupola or dome. Near this dome was a rotating black "trident" shaped device.

In the middle of the bottom of the disc was a round black spot. This was surrounded by 3 equally spaced, but smaller, pie shaped spots.

On the edge of the disc, sticking out somewhat, was a cylindrical, tailpipe shaped object which was the source of the flashes of light seen previously. This cylindrical object had emitted different shades of red and yellow light during the sighting.

The crew also noted that in the sunlight the object shone like a "steel blade".

Case #: 5
Report Type: Nocturnal Lights
Location: Danbury, Connecticut
Witnesses: Police officers
Date: July 12, 1984

Fairfield Police Lieutenant Kevin Barry was on duty around 10:40 PM on July 12, 1984. He was driving north on Route 7, near Danbury, when he noticed bright white lights in the sky to the left of his patrol car. There were numerous lights grouped in about 9 or 10 clusters.

But what really impressed officer Barry was the size of the object judging by the location of the various lights. He thought it was about the size of a football field.

Officer Barry pulled into a side street to get a better view of the lights. In the 10 or so seconds he took to complete this maneuver the lights moved from directly overhead to another part of the city some 2 to 3 miles away.

On the same night that Lieutenant Barry saw the UFO, his Chief of Police, Nelson Macedo, also had an unusual sighting. Chief Macedo was out fishing on Candlewood Lake with his 15 year old nephew and two adult men around 11 PM on July 12, 1984. The chief's nephew suddenly drew everyone's attention to a large, dull grey, circular craft high above the lake. It had twenty to thirty lights of different colors including red, green, orange and blue. The lights appeared to be turning in a circular motion while the object remained stationary.

When Chief Macedo turned off the boat lights the object responded by shutting off its lights! The Chief then tried turning the boat lights on and off for awhile and again the object responded but this time by turning its lights on brighter and brighter until they were almost yellow in color.

Finally the object moved away over a nearby mountain. The chief was unable to explain what he saw that night. He later found out that 12 of his officers also saw UFOs on the same night as his sighting.

Case #: 6
Report Type: CE-II
Location: Peekskill, New York
Witnesses: Security Guards
Date: July 24, 1984

This sighting was part of the series of sightings that lasted for more than 3 years in the Hudson Valley area of southern New York State. The witnesses in this case were highly credible security guards at a nuclear power plant on the Hudson River.

On the night of July 24, 1984 security guards at the Indian Point nuclear reactor complex in Peekskill, N.Y., observed strange lights in the sky approaching the power plant. They had previously seen similar lights on June 14, 1984 which turned out to be a huge UFO that had hovered over the plant on that night. Needless to say the reappearance of the lights caused considerable excitement at the plant.

Five of the security guards went outside the power plant to watch the lights. The most noticeable lights were arranged in a semicircular pattern and would alternately change from all yellow, to all white and then to all blue. Further back from these lights was a single, blinking red light.

The lights were clearly attached to a much larger object since the stars were blocked out over a large area around the lights. One of the guards estimated that the object was about the size of three football fields! It came within 500 feet of the guards as it moved slowly over the complex. Interestingly, the object came within just 30 feet of the only nuclear reactor of the three at the complex that was operating that night.

At one point the object moved directly over the awed guards. When it was overhead they noticed two dark, apparently hollow, spheres on the underside of the craft. These dark areas are often observed on the bottom surface of UFOs. For example in case number 4 in this chapter the Russian crewmen noticed a dark, circular area in the bottom center of the UFO that followed their ship. This dark area is often reported as remaining stationary while the rest of the disc shaped UFO oscillates. As an educated guess I would assume that this area is an observation port for the UFO occupants. Since the crew quarters of UFOs seem to be at the center of the craft, this would seem reasonable.

The plants security system failed completely as the object approached the complex. Motion detectors and alarm systems simply shut down.

On the same night as the Indian Point power plant sightings, Bob Pozzuoli of Brewster, N.Y. videotaped a disc shaped object with six lights on it traveling across the sky. Brewster is about twenty miles from Indian Point.

Also on the night of the 24th there were numerous other sightings of UFOs in the Peekskill area.

Case #: 7
Report Type: Daylight Disc
Location: Near Darmstadt, Germany
Witness: Carolyn Klingelhoeffer
Date: Fall, 1986

In July of 1997 I met Carolyn Klingelhoeffer on a hike in the White Mountains of New Hampshire. Besides being great company on the hike Carolyn also provided me with a first person account of a UFO she had observed while living in Germany in the 1980's. Her account is in the form of a letter she wrote to her mother back in the United States at the time of the sighting.

One has to wonder sometimes how many good UFO sightings remain unreported unless the witnesses meet someone like myself who has a serious interest in the topic. As we hiked up the Crawford Path toward the AMC's (Appalachian Mountain Club) Mizpah Springs hut in the Presidential Range I listened with fascination to Carolyn's UFO report. Because she had documented her sighting right after it occurred, in the letter to her mother, I knew this sighting would make an excellent addition to my collection of UFO reports. Carolyn was kind enough to allow me to publish excerpts from her letter, which follows.

"23 September 1986
"Darmstadt, Germany

"Hi Mom,

"By the time you get this letter you will probably have heard about the UFO sightings in Germany and Luxembourg. Well, it's no joke, because I saw the 'UFO' myself this morning on the way to work. The skies weren't at all clear but rather misty, so that even the sun was covered in a mist and not clear. It was about 7:40 and I was at the crossing where I turn to Kelley Barracks (near Wertkauf). (Driving north and taking a left turn toward the west.) As I was waiting at the light, this weird 'thing' floated across the sky. The front of it was shaped like an elongated egg and glowed white. A ribbony band was attached and trailed behind it and glowed a translucent purple and green. At the end was a smaller white glowing point. It drifted through the air above me at a good speed, about like a fast plane but not like the fighter jets around here, not so fast. When the light turned green and I drove around the corner, it had disappeared. I don't think it could have been very high, because it was so hazy, that it wouldn't have been visible. It's difficult to estimate, but I think it was about as high as the planes fly around here that start and land in Frankfurt, and it was probably as long as a Jumbo. Anyway, after seeing this, I did a double-take. It all went by so fast.

"Several hours later I heard the news on the radio, and apparently, reports had started coming in about sightings around Stuttgart and Karlsruhe and then in the Frankfurt-Darmstadt area up to the Rheinland-Pfalz and Luxembourg. By the end of the day, they were interviewing eye-witnesses, among them pilots flying near Frankfurt and an air-controller in Luxembourg. Hundreds of people were calling the police and airport. Many people including the air-controller, spotted at least 5 of these 'things.' I just watched the news on TV, and I've been listening to the radio all day, but up to now, no one has a plausible answer. The so-called 'scientists' say it must have been a meteor entering the atmosphere. This 'UFO' glowed, it didn't burn, and it didn't fall, it kept on going, just like a plane. Others say, pieces of old satellites are falling back to earth, but these too would fall like a meteor. And the best explanation came this evening that it was only a reflection of a plane in the sunlight. But in so many different places and so many different and numerous reflections? Another point that has been played down is that the Bundeswehr and the US military are now involved in maneuvers. They disclaim any knowledge of any mishaps or that the military is involved in any way. 'Others' say that perhaps a rocket or several rockets have been inadvertently released, that is, by mistake or malfunction. Who knows? Anyway, its been a creepy day. All I know is that I saw this 'UFO' and its comforting to know I wasn't alone......."

Case #: 8
Report Type: CE-I
Location: Eupen, Belgium
Witnesses: Family members and neighbor
Date: December 1, 1989

This sighting was part of a wave of sightings that occurred in Belgium between November 1989 and June of 1990. Auguste Meessen, a physicist from the Catholic University of Louvain, investigated this sighting along with a number of other sightings in Belgium and neighboring Germany. The report itself appeared in the May/June 1991 issue of the International UFO Reporter published by the J. Allen Hynek Center for UFO Studies.

The father of five children was sitting near a window in his house on the evening of December 1, 1989. It was 6:50 PM and two of his five children, aged 14 and 15, were playing in a courtyard outside the window with another youngster from next door. One of his children excitedly called his attention to an object in the sky that the youngsters were looking at. Going out to the courtyard the father looked up into the sky and almost fell over backwards at the sight of an enormous lozenge-shaped object almost directly overhead! The lozenge-shaped object had two round white lights at each of its four corners and at the center an upside down dome or gondola like structure hung below (again we see this apparent observation area for the UFO occupants). The white lights at the corners were very powerful and illuminated the ground with brilliant white light. The gondola structure was bathed in orange light and at its bottom was a dark green light. Near the top of the gondola structure where it connected to the lozenge-shaped part of the craft, there was a row of pulsating red lights that surrounded the gondola.

The maximum width of the object between two opposite corners was about 115 feet. The father estimated this based on the number of neighboring houses the object covered. The lozenge-shaped part of the object was dark but clearly visible against the star-lit sky. It gradually disappeared in the distance.

The father had a second sighting on January 10, 1990 at around 7:35 P.M. while driving on a highway. It appeared to be exactly the same object but this time it was hovering stationary in the sky. He pulled his car off the road and got out to watch the object. It stayed motionless for between 10 to 15 minutes and then began to move away. Of particular interest is that as the object started to move its luminosity suddenly increased. This is an often reported characteristic of UFOs that as they accelerate or move from a stationary position the luminosity increases significantly.

Case #: 9
Report type: Radar Visual
Location: Belgium
Witnesses: Belgian Air Force pilots, police, radar operators and others.
Date: March 30, 1990

Belgian authorities, both military and civilian, are apparently much less secretive about UFO sightings than their counterparts in many other countries. This next report was also part of the now famous Belgian UFO sighting wave. It contains some excellent military radar data.

On the night of March 30, 1990 the Belgian national police reported a UFO sighting and this was confirmed by two separate ground radar installations. One radar was part of the NATO network and located southeast of Brussels in the city of Glass while the other was a Belgian radar at Semmerzake used to control both civilian and military air traffic over Belgium. The two radars covered a range of 300 kilometers.

Around midnight two F-16 fighters were scrambled and ordered to locate the UFO. Both F-16's picked up the UFO on their sophisticated radar sets and locked on to the target. But the lock-on only lasted about six seconds. The target, which had been moving at 280 KPH suddenly accelerated to 1,800 KPH, descending from 3,000 meters to 1,700 meters in just one second! Moments later the object swiftly descended another 1,500 meters and was lost to both the ground radars and the on board radars of the F-16's.

This remarkable game of cat and mouse was witnessed from the ground by 20 national police and numerous other witnesses. The ground observers reported seeing both the F-16's and the UFO at the same time. The total encounter lasted some 75 minutes. Despite the extreme acceleration of the UFO, no sonic boom was heard.

Case #: 10
Report type: CE-II
Location: Belgium
Witnesses: Husband and wife
Date: May 4, 1990

The Belgian wave of UFO sightings produced a number of sightings of rectangular and triangular shaped objects. But there were also other shapes reported such as discs and cones. This sighting was of a cone shaped object seen at close range.

The primary witness was a 72 year old archeologist living in the village of Stockay, Belgium. On the evening of May 4, 1990 he was closing the door to his greenhouse in the garden when his cat suddenly nestled itself between his legs. At about the same time the dogs in the neighborhood began howling in an unusual manner. On his way back to his house he was startled to see a luminous object in a field less than 200 meters away.

He attempted to alert his neighbors but there was no response. He then persuaded his wife to come out to see the strange object. Standing in front of their house they could see a luminous, cone shaped object the top of which resembled a mushroom. After watching it for 5 minutes the archeologist's wife decided to go back indoors.

The archeologist then decided to get closer to the object and walked through the field till he was about 100 to 150 feet from the unknown craft. At this distance he got a good look at the structure of the object. It was approximately 8 to 10 meters at the base and perhaps 4 to 5 meters high. The lower part was luminous and opaque while the center part was white with yellow edges. The light emitted by the object was intense.

As he watched the object the mushroom shaped top section separated from the lower section and rose up in the air, turning a brilliant orange color in the process. Rather frightened by this point the archeologist retreated some dozen paces from the object but continued to watch from a safer distance.

Finally he returned to his house and joined his wife on the front landing, watching the object for a few more minutes before going back inside.

Later investigation showed definite marks on the ground where the object had been. At the landing site there were four circular impressions in a rectangle of 8 by 10 meters. The grass was twisted and flattened at the center of each circle. A fine yellowish powder coated the grass blades at the site.

Case #: 11
Report Type: Daylight Disc
Location: Sighting from aircraft near Paris, France
Witness: Crew of Air France Flight
Date: January 28, 1994

The primary witnesses in this January 28, 1994 sighting were Air France Captain Jean-Charles Duboc along with his flight crew. Captain Duboc and his crew observed a large, reddish disc shaped object that became transparent before their eyes. On November 12, 2007 at the Washington, D.C. National Press Club Conference on UFOs organized by documentary film producer James Fox, Captain Duboc gave the following testimony:

"My name is Jean-Charles DUBOC, and I am a retired Air France Captain.
During Air France Flight 3532 from Nice to London, on 28 January 1994, 1 observed, with my crew, a dayliqht UFO near Paris. We had few passengers on board and none of them had reported that observation to the crew.

The time was 1:00 P.M, and the visibility was excellent with some altocumulus clouds.

That object had been identified initially by a steward who was in the cockpit and by the copilot as a weather balloon, but when I identified it, it looked quite different.

In fact this UFO was in evolution and it looked like this (I shall then show a drawing), with a bank angle of 45°. It seemed to be a huge flying DISC. It stabilized and stopped moving (moving horizontally the draw.

We observed it over one minute on the LEFT on our plane, surprisingly totally stationary in the sky, and it disappeared progressively (turn of the drawing).

This large object was below us at the altitude of 35, 000 ft (we were at 39, 000 if), at a distance of about 25 nautical miles.

The colour was RED BROWN with BLURRED outlines.

The apparent diameter of this object could be compared to the diameter of the moon, or the sun. That means that it was about 1000 feet wide.

We had no idea of the structure of the UFO that seemed to be embedded in a kind of magnetic or gravitational field, with no lights or visual metallic structure, which gave it a really FUZZY appearance.

The most incredible aspect is that it became transparent and disappeared in about 10 to 20 seconds. My co-pilot, chief steward and I quickly realized that what we were seeing did not resemble anything known to us, and we reported our sighting to Reims traffic control. Simultaneously, the radar of the Operational Centre of the Air Defence (the CODA), registered a one minute spot crossing the track of our plane. When I recorded the estimated position of the UFO on an aeronautical map I was surprised to realize that its position was near the base of Tavemy which hosts the headquarters of the French Strategic Air Command.

This sighting has been studied by the French military COMETA group under the direction of General Denis Letty and with the participation of several high ranking officers of the French Defence, as well as by GEPAN under the French National Space Centre (CNES I.

The investigations determined this could not have been a weather balloon, and estimated the approximate length of the UFO to have been 800 feet.

This sighting remains unexplained."

UFOs that become partially then fully transparent in front of witnesses are relatively rare events.   Yet this type of sighting I believe is perhaps the most important clue of all to the physics of UFOs.   James Fox's DVD on this and other UFO events is a valuable addition to the study of this phenomenon.

Case #: 12
Report Type: Nocturnal lights
Location: Arizona
Witnesses: Estimated at over 10,000
Date: March13, 1997

This particular sighting of glowing orbs and triangular objects has come to be known as the "Phoenix Lights" and is one of the best documented events in the history of UFO reports. Although the March 13, 1997 observations were the centerpiece of this event, UFO sightings in and around Arizona have in fact been occurring since at least 1995 and have continued right through the first decade of the 21st century. One of the ongoing mysteries of the UFO phenomenon is why these craft seem to repeatedly visit a given region over extended periods of time. The Hudson Valley UFO sightings, for example, took place from 1982 to 1995.

Dr. Lynne Kitei, a physician who lives with her family just outside the city of Phoenix, has done a superb job of documenting and researching the series of sightings that comprise the Phoenix Lights. From her mountainside home Dr. Kitei was in a perfect location to view and photograph many of the orbs and triangles that appeared around the Phoenix area. One of her pictures can be seen in the picture section of this book.

The brief summary of the Phoenix Lights included here does not do justice to these extraordinary events. Dr. Kitei's books, DVD's and website on these UFO sightings in Arizona provide an excellent and comprehensive study of this phenomenon for anyone with a serious interest in the UFO enigma.

It is estimated that over 10,000 people in Arizona observed the orbs and triangles seen on the night of March 13, 1997. The triangular object (or objects) was observed directly overhead by a number of witnesses as it traveled at a modest speed from northern Arizona all the way down to Phoenix and Tuscon. It appeared to be at a relatively low altitude of below 3,000 feet, was possibly 1,500 feet in width or more, seemed to have a dark metallic structure and had lights at each corner. Other witnesses observing the same or possibly different V shaped craft reported that there were multiple lights along the "wings". Some of those who observed the V shaped craft reported that the space between the lights on the wings gave the appearance when looking at the stars above as if you were looking through water. This, I believe, is a significant observation that has also been reported in some other UFO sightings around the world. As we shall see this apparent partial transparency is a vital clue to UFO technology.

The glowing orbs are smaller objects, perhaps only a few feet in diameter. Since they sometimes appeared near the ground it is possible that these are probes used for close observation similar to the famous foo balls of World War II that were seen by airmen of all nations. Some observers also reported that the orbs looked like they were swirling or rotating. This is not unexpected since the large disc shaped UFOs are likewise frequently reported as having part or all of their structure rotating. This is another important clue as to how UFOs function.

Case #: 13
Report Type: Daylight Disc
Location: Chicago O'Hare International Airport
Witness: Pilots, aircrew, airline ground personnel and others
Date: November 7, 2006

Credible UFO sightings have continued to be reported in the first decade of the 21st century. On November 7, 2006 a disc shaped UFO was reported in the vicinity of Chicago's O'Hare Airport. It appears to have first been seen by the pilot and crew members of a commercial airliner at around 3:45 PM on November 7. The witnesses observed a cigar or disc-shaped object while cruising at 37,000 feet about ten minutes after they departed O'Hare. The pilot was concerned that the unidentified craft was "dangerously" close to his aircraft. A flight attendant took several photographs of the UFO which can be seen on YouTube.

Probably the same UFO was observed again at 4:30 PM hovering above Gate C17 at O'Hare. Witnesses on the ground included at least one pilot in the cockpit of his aircraft and a number of other airline employees.   The UFO was described as disc shaped, dark gray in color and clearly visible.   After hovering several minutes it ascended rapidly up into and through the low cloud ceiling at 1,900 feet.   The witnesses reported that the object left an "eerie" hole in the cloud cover.

Case #: 14
Report Type: Daylight Disc
Location: English Channel near Channel Islands
Witness: Pilots and passengers
Date: April 23, 2007

Occasionally UFO's of enormous size are reported by credible observers.   In this case Captain Ray Bowyer of Aurigny Airlines sighted an object that could have been as much as a mile in width. The Japan Airlines UFO sighting in 1986, included in the Classic UFO Cases in Section I, estimated the width of that craft to be about 2,000 feet.   The Phoenix Lights craft discussed above was also of great size.

Captain Bowyer was approaching the island of Guernsey in the English Channel when he observed a stationary flattened disc shaped object of tremendous size that glowed a brilliant yellow color.   The craft had a dark green area on one side.   Moments later Captain Bowyer spotted an identical craft that was some distance behind the first object.   The UFOs were under observation for around 15 minutes.   Some of the passengers witnessed the strange craft as well.   Another pilot with Blue Lines Airline flying nearby also reported sighting at least one of the objects.

Skeptics suggested that these sightings were due to light beams from a lighthouse reflecting off lenticular clouds.   However this is unlikely since the objects were under observation for an extended period of time which would change the viewing angle and alter the nature of the reflection.   Furthermore there were two identical objects which would require two identical lenticular clouds!   It is also unlikely that these objects would have been detected by weather satellites since they were at a relatively low altitude beneath or between cloud layers.

The brilliant luminosity of these craft is often reported in UFO sightings and, in my opinion, is an important characteristic that is giving us clues to the technology behind this phenomenon.

Salem, Massachusetts 1952, Coast Guard

This is an excellent photo showing the often reported intense self luminosity of UFOs. It was taken by United States Coast Guard seaman Shel Alpert at the Salem, Massachusetts Coast Guard Station on July 16, 1952. The brilliant luminosity of UFOs provides an important clue as to how they function. **Photo source: ufocasebook.com.**

Namur, Belgium, 1955 photographer unknown

Here is another interesting example of the self luminosity of a UFO. In this case only the bottom half of the object is lit up. In order for these machines to stay levitated in earth's gravity it makes sense that the "action" is at least sometimes located just on the lower surface. This June 5, 1955 photo was taken in Namur, Belgium. **Photo source: ufocasebook.com.**

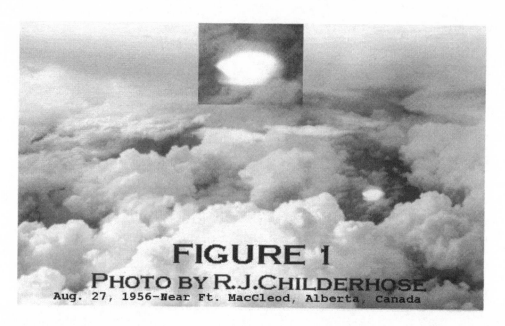

FIGURE 1
PHOTO BY R.J.CHILDERHOSE
Aug. 27, 1956-Near Ft. MacCleod, Alberta, Canada

On the evening of August 27, 1956 Royal Canadian Air Force F-86 pilot R. J. Childerhose took this picture of a very bright disc shaped UFO.   He was flying at 36,000 feet near Fort MacCleod, Alberta around 20 minutes before sunset.   Optical physicist Dr. Bruce Maccabee estimated the energy output of the object was in excess of a gigawatt based on the spectrum range of the film. This tremendous energy output is another important clue to the technology of UFOs.   **Photo source: ufocasebook.com.**   **© Childerhose.**

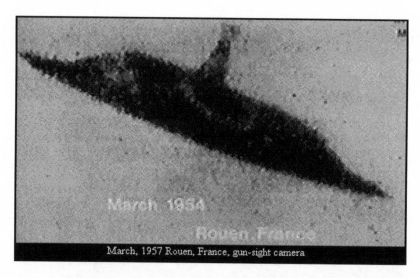

March, 1957 Rouen, France, gun-sight camera

UFO sightings occur all over the planet.   In March of 1957 a French fighter pilot took this gun camera image of a UFO over the city of Rouen.   The object stayed with the aircraft for a short time and then accelerated to a speed beyond the performance capability of the fighter.   These pictures and experiences from military sources lend credibility to the assertion that we are very likely witnessing craft of extraterrestrial origin.   **Photo source: ufocasbook.com.**

INSERT

Edwards A F B 1957

This is another picture of a UFO taken by the military, this time by the U.S. Air Force in September, 1957 near Edwards Air Force Base in California.   The UFO was apparently observing the B-57 bomber.   **Photo source: ufocasebook.com.**

Tiopati Lake, New York, 12-18-1966

In December, 1966 Vincent Perna took this photo of a UFO from the shoreline of Lake Tiopati Reservoir in Bear Mountain State Park, New York.   So called UFO "waves" occurred in 1952, 1957 and 1966 according to Project Blue Book data.   Dr. Hynek and NICAP both felt that the photo was genuine.   This looks like the often reported daylight disc but seen from the side.
Photo source: ufocasebook.com.   © Michael Hesemann.

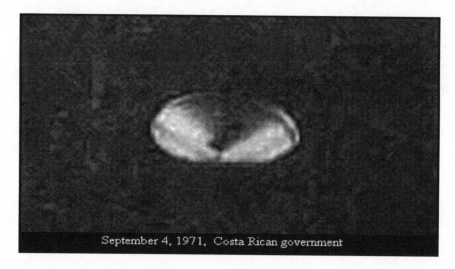

September 4, 1971, Costa Rican government

A mapping aircraft photographed this typical metallic disc shaped UFO over the Arenal region of Costa Rica on September 4, 1971.   The object in this picture was thoroughly analyzed but remains unidentified   One of the recurring features of most UFO reports is the observation that at least some part of the UFO has a circular configuration whether its overall shape is a disc, cylinder or sphere.   Even triangle UFOs appear to have large brightly lit circular areas at each corner.   As we will see in Section III of this book the circular configuration aspect of UFOs is another important piece of data relating to the "engine" that powers these craft.   **Photo source: ufocasebook.com. Original photo from the Government of Costa Rica.**

1976-Amazon Jungle, Brazil

This is a remarkable cockpit photo of a UFO taken by the pilot of a Brazilian Airlines Boeing 727 in 1976.   UFOs are often reported to show an interest in aircraft but relatively few airborne sightings are accompanied by a photograph.   In many close encounters with aircraft UFOs exhibit inertia defying maneuvers that must be taken into account in any attempt to explain their technology.
Photo source: ufocasbook.com.   © Michael Hesemann.

Vancouver, Canada-October 8, 1981 Hannah Roberts

The above photograph of a disc shaped object was thoroughly investigated by Dr. Richard F. Haines and included in a 1997 major study of the UFO phenomenon commissioned by Laurance S. Rockefeller and led by Stanford University astrophysicist Dr. Peter Sturrock.   The photo was taken by Hannah Roberts on October 8, 1981 at Vancouver Island, Canada.   Dr. Haines, a former NASA scientist, concluded that the photograph does represent an unidentified aerial disk-like object.
Photo source: ufocasebook.com.

Petit Rechain, Belgium, 1990.

During 1989 and 1990 there was another classic wave of UFO sightings, this time concentrated in Belgium. The photo above was taken in Petit-Rechain, Belgium in April of 1990. Triangle shaped UFOs with bright circles of light at each corner and usually a pulsating red light in the center figured prominently during this wave. **Photo source: ufocasebook.com.**

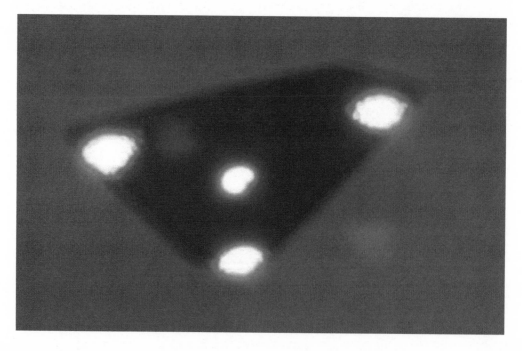

This is another example of the triangle shaped UFOs seen over Belgium between 1989 and 1990. It was snapped on June 15, 1990 by J.S. Henrardi in Wallonia, Belgium. This photo also appears on the cover of this book. Note the bright lights at each corner. UFOs of different shapes and sizes have been observed over the years and may indicate we are being visited by a number of different extraterrestrial civilizations. **Photo source: ufocasebook.com. The photographer has released this photo into the public domain.**

Three unexplained amber orbs over Phoenix.
10.10.01 (One month after 9/11)

On the evening of March 13, 1997 a series of UFO sightings occurred in the state of Arizona that climaxed in the Phoenix area and were witnessed by over 10,000 people.   These UFO reports were so impressive that they made the national and international news.   In fact sightings of unidentified flying objects took place both before and after the March 13, 1997 events and have come to be known as the "Phoenix Lights".   The above picture of three unidentified orbs over Phoenix was taken by Lynne D. Kitei, M.D. October 10, 2001, just one month after 9/11.   Dr. Kitei has done an outstanding job researching and documenting the Phoenix Lights.
© 10.10.01, Lynne D. Kitei, M.D.

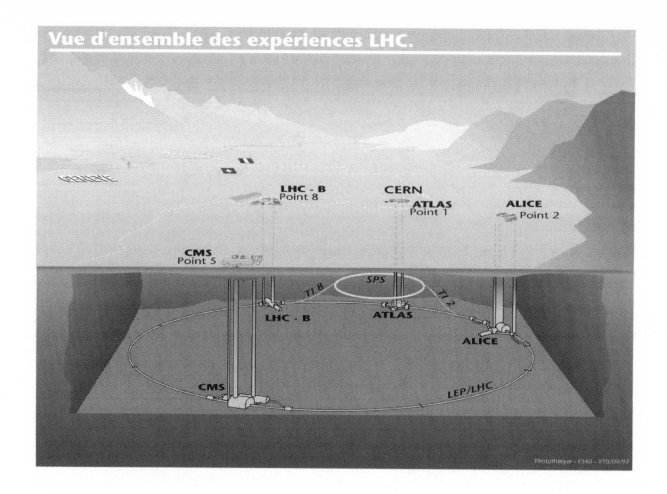

The CERN Large Hadron Collider (LHC) illustrated above is the biggest particle accelerator ever built. Twenty nations are funding this multi-billion dollar device that is located on the Swiss and French border near Geneva. It essentially consists of two pipes buried deep underground that will carry subatomic particles of the hadron family in opposite directions and then collide them at detectors located around the rim of the 17 mile in diameter circular track. Giant super cooled magnets will be used to accelerate and guide the subatomic particle beams at near light speed around the LHC circular pipe track. At the detectors scientists will be looking for evidence of extra dimensions, the Higgs particle, what the universe looked like in the first microseconds after the Big Bang, KK gravitons, dark matter, dark energy, etc. **© CERN**

The ATLAS Detector pictured above is one of six detectors located around the LHC ring.   Note the immense size of the detector by comparing it to the worker who is located at the front bottom of the picture.   **© CERN**

Here is a view of a segment of the LHC tunnel showing the casing that encloses the pipes within which the particle beams travel.   The LHC ring is buried hundreds of feet underground.   **© CERN**

One of many groups of school kids who have been visiting the LHC site.   The expected life of the particle accelerator is about 15 years so some of these children might someday work at the LHC.
© CERN

The author with daughter Debbie photographed in Italy during a June, 2010 Mediterranean vacation.

The author with wife Carol photographed at the Roman Forum, June, 2010.

Section II

The Evolution of Modern Physics

Classical Physics Before the 20th Century
Chapter 1 in Section II

Newton's Laws of Mechanics and Gravity

Without question the English physicist Isaac Newton is responsible for laying the foundations of modern physics.   Newton lived from 1642 to 1727 and is credited with discovering the basic laws of mechanics and gravity as well as being the co-inventor, along with Gottfried Leibniz, of the branch of mathematics known as calculus.   Calculus was essential to Newton's determination of his three laws of motion.   These laws are described briefly below:

First Law - An object remains at rest or moves in constant motion in a straight line unless acted upon by a force.

Second Law – Newton's second law says that when a force is applied to a body it changes the momentum of that body in a manner that can be described exactly by a mathematical formula.   Momentum is defined as $p=mv$ where "**p**" is momentum, "**m**" equals the mass of an object and "**v**" is the velocity of the object. The force on a body is measured by the formula $f=ma$ where "**f**" equals the force, "**m**" equals the mass and "**a**" is the acceleration.

Third Law – The third law states that whenever a force is applied to an object, the object reacts with an equal and opposite force.

Its important to note that in Newton's second law that the mass "**m**" in the equation $f=ma$ is the *inertial* mass of an object as opposed to its gravitational mass.   That is, it would take the same force to accelerate an object to a given speed of mass "**m**" in the weightlessness of outer space as it would on the surface of the earth (assuming no friction).   It doesn't matter if the object is in a gravity field or not.   The inertia of an object can be thought of as the resistance of an object to changes (ie. accelerations) in its uniform motion.   Its inertial mass is a reflection of this resistance.

It is of particular interest that the inertial mass (i.e. $m=f/a$) of an object has been found to be exactly equal to the objects gravitational mass when that object is in a gravitational field.   This has never been entirely understood or explained with certainty.   However Ernst Mach, an Austrian physicist, in the 1860's suggested that the inertial mass of an object is due to the effect of all the masses in the entire universe acting on the object.   This is known as Mach's principle.   In some way all the distant galaxies and stars contribute to the inertia, or resistance to change in motion, of an object on earth.   How this might happen is not known but Einstein tried to incorporate Mach's principle in his theory of general relativity but at a later date remained unsure if he had accomplished this.

I mention this important puzzle since it may be significant in explaining the physics of the UFO phenomenon.   Interestingly the key step toward the development of Einstein's general theory of relativity was his equivalence principle that assumes gravitational mass and inertial mass are equivalent.   Yet to this day no one knows why the equivalence principle holds true.   Later in

Section III I'll propose a possible explanation for the equivalence principle and Mach's principle based on some of the latest cutting edge theories being developed in the physics community.

Besides the laws of motion, Newton is also credited with the first mathematical formula to measure the force of gravity and another formula for measuring kinetic energy. These equations are some of the most fundamental formulas of physics and are instrumental in deriving other key math relationships in modern physics.

Newton's formula for calculating the gravitational force between two masses is as follows:

$$F = \frac{Gm_1m_2}{d^2}$$

$F =$ **Force of attraction between two masses. An example might be a cannon ball sitting on the earth's surface and the earth itself.**

$G =$ **Gravitational constant. This is a mathematical constant that Newton derived through experiment but it is a universal constant that holds for all masses everywhere.**

$m_1 =$ **The mass of one of the two objects.**

$m_2 =$ **The mass of the second object.**

$d =$ **The d in the d^2 is the distance between the centers of the two masses.**

With Newton's formula for calculating the gravitational force it is possible to determine virtually all the math needed for orbiting satellites in space, the orbits of the planets around the sun as well as many other examples.

However in the 1800's scientists did discover a tiny discrepancy in the orbit of the planet Mercury that could not be explained by Newton's theory of gravity. It would take Einstein's theory of general relativity to ultimately explain this anomaly.

Another key formula describing the physical world that was developed by Newton is his equation that determines the kinetic energy of an object. The kinetic energy is the energy of motion of an object as shown below:

$$KE = \frac{1}{2}mv^2$$

$KE =$ kinetic energy

$m =$ mass of object

$v =$ velocity of object

Maxwell Unites Electricity, Magnetism and Light

In the 18th and 19th century scientists began unraveling the mysteries of electricity, magnetism and light.   It was found, for instance, that the force between charged particles diminished by the inverse square of the distance between the charges.   This was also found to be true for magnetic poles when they are moved apart..   Scientists noted the similarity to the gravitational force between two masses which by Newton's formula also decreases by the inverse square of the distance between the masses.

By the 1850's a young Scottish scientist named James Clerk Maxwell had determined the laws governing the electric force, magnetic force and their connection to light waves.   Maxwell's laws can be summarized as follows:

1) Stationary charges produce static electric fields.
2) Stationary magnets are the source of magnetic fields
3) A fluctuating magnetic field creates a changing electric field
4) An electric current and/or a fluctuating electric field creates a changing magnetic field

The actual equations Maxwell derived are these:

1) $$\nabla \cdot E = \frac{\rho}{\varepsilon_0}$$

2) $$\nabla \cdot B = 0$$

3) $$\nabla X E = -\frac{\partial B}{\partial t}$$

4) $$\nabla X B = \mu_0 J + \mu_0 \varepsilon_0 \frac{\partial E}{\partial t}$$

Don't let these equations scare you! They really aren't all that hard to understand. They use a branch of *calculus* called *vector calculus* that was developed in the 19th century. In Appendix A, I explain basic calculus followed by a brief review of vector calculus and then an explanation of Maxwell's laws. Understanding vector calculus and how it is used in Maxwell's equations isn't necessary for this book but I included it in the appendix for interested or curious readers. But don't be afraid to check out Appendix A. Calculus is actually pretty easy to understand. Since math is the language of science the goal of Appendix A is just to give an idea of the concepts involved, understand the funny looking symbols and take the mystery out of science! So check it out if you'd like.

After Maxwell completed his equations describing electromagnetism he discovered something amazing. By combining equation 3 with a modified version of equation 4 Maxwell discovered that a changing magnetic field creates a changing electric field. And a changing electric field creates a changing magnetic field. Maxwell soon realized that this created a self sustaining electromagnetic wave which is the source of visible light as well as other types of radiation like X-rays, infrared, ultraviolet, etc. He also calculated that his equations required this electromagnetic radiation to travel at a constant speed of about 186,000 miles per second. It would take nearly half a century before the full significance of the constant speed of light would be understood and usher in a revolution in our understanding of nature. In the next chapter we'll see the consequences of Maxwell's laws.

Special Relativity
Chapter 2 in Section II

Introduction

Einstein's 1905 theory of special relativity is much easier to understand than most people realize. The theory arose out of a need to explain the puzzling fact that the speed of light does not seem to depend on the motion of the observer. If you are in a moving boat and water waves approach your vessel from behind you will notice these waves appear to go slower when they pass your boat as your boat goes faster. But if you turn the boat around and head opposite the direction of the waves you'll notice the waves go by your boat faster as your boat goes faster. Since light was known to consist of waves which were thought to be moving through an invisible "ether", it seemed to scientists that light should travel slower as you move in the same direction of a beam of light and faster as you move opposite into a beam of light. Just like water waves. Unfortunately numerous experiments proved that light waves always traveled at the same speed regardless of whether you were moving with or against the beam of light. Several scientists had come up with different ideas as to how to explain this puzzling fact but it was the young German-Swiss patent office clerk Albert Einstein who put it all together.

Time Dilation

In essence the special theory of relativity says the laws of physics are the same in all inertial frames. Inertial frames are any reference frames that move at constant velocity relative to each other. By the laws of physics Einstein meant both Newton's laws as well as Maxwell's laws of electromagnetism which represented the core of what was known about physics in 1905 when special relativity was formulated.

Since Maxwell's laws require that light always travel at a constant velocity of 186,000 miles per second or "c" then it follows that all reference frames will see any light beam moving at the same speed. "c" is the letter physicists use to designate the speed of light. But this means that two observers in two different inertial reference frames moving at different speeds relative to each other will nevertheless measure the same light beam that both observe independently as traveling at the known constant velocity of light "c". This is where special relativity gets interesting!

If both inertial observers moving at different but constant speeds relative to each other see the same beam of light going at the same speed then something else has to give in order for this to happen. It turns out that both *space* and *time* must be *variable* in order for the speed of light to be constant to all inertial observers! As astounding as this may seem. The following example explains exactly how space and time vary and shows how the interesting math is derived. The math is surprisingly easy so no need to worry.

-Fred and Alice

Imagine two inertial observers named Fred and Alice. Fred is standing at a railroad station observing a train whiz by from left to right at constant speed "v" with Alice onboard as a passenger. Alice has a type of clock in her railroad car called a light clock that measures time by bouncing a light beam back and forth between a mirror on the floor and another mirror directly above on the ceiling.   Since the sides of Alice's railroad car are glass this means Fred, standing on the station platform, can also observe Alice's light clock in action.   A complete cycle of Alice's clock consists of a light pulse starting at the bottom mirror, reflecting off the top mirror and returning to the bottom mirror.   This is one "click" of Alice's clock.   Fred also has an identical light mirror clock next to him on the railroad station platform.   (See illustration A).

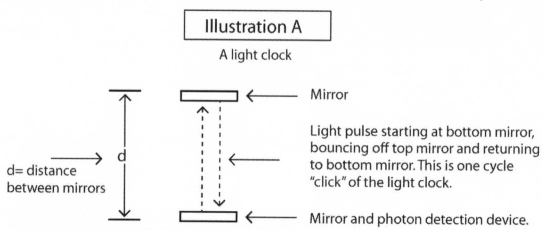

The question is: Will Fred and Alice's clocks measure the same time?   Fred knows the time for a single cycle of his clock can be calculated by simply dividing the *distance* his clock's light pulse goes in one cycle by the speed of light.   So if the vertical distance between the mirrors is "*d*" and the speed of light is "*c*" then one cycle of Fred's clock from his point of view is simply *2d/c* (i.e. a complete cycle is the light pulse going back and forth between the mirrors and then distance divided by speed gives time elapsed).

Next Fred has to measure Alice's clock for one complete cycle to compare her time to his. Unfortunately measuring Alice's cycle time isn't as easy as measuring Fred's own clocks cycle time.   Since Alice is moving at constant velocity *v*, then from Fred's point of view as Alice speeds by the light pulse from her bottom mirror does not go vertically up to hit her top mirror.   Because Alice is moving this means Fred see's her light pulse move to the right as it travels toward the mirror on the ceiling.   This is illustrated below:

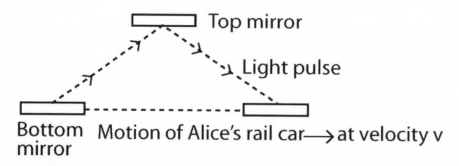

*[Alice's railroad car is moving at velocity **v** to the right as seen from Fred's point of view. By the time the light pulse starting at the bottom mirror arrives at the top mirror the rail car has*

moved to the right from Fred's frame of reference so the pulse has traveled at an angle to the top mirror where it bounces back at the same angle and returns to the bottom mirror.]

In order to compare Fred's time to what he sees as Alice's time we need to calculate the time it must take between clicks for Alice's clock versus Fred's clock.   We can do this by using the Pythagorean theorem.

If we draw a line vertically down from where Alice's light pulse bounces off the top mirror (from Fred's point of view) we create two back to back right triangles as shown in illustration C:

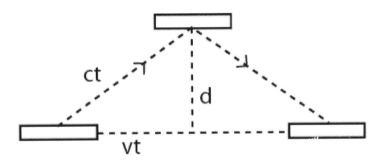

*[A complete cycle of Alice's light clock as seen by Fred on the station platform.   The cycle forms two right triangles whose sides are **ct** (**c**=speed of light, **t**=time for light beam to go from bottom mirror to top mirror), **d** (known vertical distance between top and bottom mirrors) and **vt** (**v**=velocity of the train and **t** is the same as above)]*

Knowing the definition of the distances of the three sides of the triangle we can simply use the Pythagorean theorem to come up with a formula that will give us Alice's time versus Fred's time for one cycle of each of their clocks.

From Pythagoreas we know that:

$$c^2 t^2 = v^2 t^2 + d^2$$

then
$$c^2t^2 - v^2t^2 = d^2$$

divide both sides by c^2
$$t^2 \cdot \frac{(c^2 - v^2)}{c^2} = \frac{d^2}{c^2}$$

then
$$\frac{t^2c^2}{c^2} - \frac{t^2v^2}{c^2} = \frac{d^2}{c^2}$$

simplify
$$t^2 - t^2 \cdot \frac{v^2}{c^2} = \frac{d^2}{c^2}$$

simplify
$$t^2\left(1 - \frac{v^2}{c^2}\right) = \frac{d^2}{c^2}$$

take square root of both sides
$$t \cdot \sqrt{1 - \frac{v^2}{c^2}} = \frac{d}{c}$$

rearrange

$$t = \frac{d}{c} \cdot \frac{1}{\sqrt{1 - \dfrac{v^2}{c^2}}}$$

But this is the time from Fred's point of view for Alice's light pulse to go only from the bottom mirror to the top mirror or just half a cycle.   A full cycle is two times this amount or:

$$t = \frac{2d}{c} \cdot \frac{1}{\sqrt{1 - \dfrac{v^2}{c^2}}}$$

Since we know that a full cycle of Fred's clock is *2d/c* then this means Alice's clock differs from Fred's clock by just the factor:

$$1 / \sqrt{1 - \frac{v^2}{c^2}}$$

From the factor we can easily see that the greater Alice's velocity v the longer each of her time cycles are compared to Fred's clock.   This means that for every click of Alice's clock, Fred experiences a greater number of clicks if Alice is going at a high velocity v.   For example if Alice is going at 90 percent the speed of light (this is a really high speed train!) then the formula comparing Alice's time cycle to Fred's time cycle is this:

$$\text{time factor difference} = \frac{1}{\sqrt{1 - \frac{v^2}{c^2}}}$$

The speed of light is: $c = 186{,}000$ miles/second

Alice at 9/10ths the speed of light has velocity v of:

$$v = 167{,}400 \text{ miles/second}$$

Inserting these values into our formula we get:

$$\text{time factor difference} = \frac{1}{\sqrt{1 - \frac{(167{,}000)^2}{(186{,}000)^2}}}$$

Calculating we get:

$$\text{time factor difference} = \frac{1}{\sqrt{1 - 0.81}}$$

Simplified:

$$\text{time factor difference} = \frac{1}{\sqrt{0.19}} = \frac{1}{.436} = 2.29$$

Since we multiply Alice's time factor times Fred's time formula we see that each of Alice's time cycles is 2.29 times longer than Fred's time cycle. So if each of Fred's time cycles is say one minute and he experiences 10 minutes (cycles) on his clock then Alice will only show 10/2.29=4.3668 cycles on her clock or just 4.3668 minutes. In other words time slows down for Alice the faster she goes relative to Fred. This is known as *time dilation*.

This time dilation affects not only Alice's light clock but also any other type of clock she has on board her railroad car whether it is mechanical, digital or whatever. All her clocks will run slow relative to Fred.

Time dilation has been confirmed experimentally many times. In one experiment extremely sensitive clocks were put on board jet airliners and at the end of their journey it was found that these clocks did indeed run slow compared to clocks on the ground.

Length Contraction

Let's say Alice from the previous example marks the railroad track at points P and Q to mark the beginning and end of her 4.3668 minute journey. Since she knows her velocity is v relative to Bob then she measures the distance between points P and Q as simply velocity v multiplied by time of 4.3668 minutes or $4.3668v$.

Bob however sees Alice move from point P to point Q in 10 minutes rather than the 4.3668 minutes of dilated time that Alice experiences. Therefore Bob measures the distance between points P and Q as 10 minutes multiplied times velocity v or $10v$.

Note that Bob measures Alice's speed as v relative to himself and she in turn measures her speed as v relative to Bob, just as two people riding by each other in opposite directions would measure the same relative velocity. This means v is the same for both Bob and Alice.

But clearly the $10v$ length Bob measures is greater than the $4.3668v$ length that Alice measures. So Alice from her frame of reference in the railroad car sees the distance between P and Q along the railroad tracks (which is in Bob's frame of reference) as much shorter than what he measures. The shorter length that Alice measures between points P and Q versus Bob's measurement is known as the *Lorentz-Fitzgerald contraction*. It is named after the two physicists, Henrik Lorentz and George Fitzgerald, who first predicted this strange effect of relativistic physics.

The Lorentz-Fitzgerald contraction works in both directions so Fred sees Alice's railroad car as being shorter than what Alice measures it to be. The contracted length of any object would be determined by the following formula:

$$L' = L\sqrt{1 - \frac{v^2}{c^2}}$$

where L = original length of object

And L' = shortened length

Relativistic Mass Increase

Some years after the introduction of the idea of length contraction of speeding objects Henrik Lorentz developed a theory on how the mass of an object would also be affected by its speed relative to an observer.   Lorentz proposed that the mass of a charged particle such as an electron was inversely proportional to its radius.   A particle that is compressed along the direction of its motion relative to an observer would then have an increase in its mass.   The foreshortened radius of a particle along its direction of motion is calculated from the Lorentz-Fitzgerald contraction as follows:

$$R' = R\sqrt{1 - \frac{v^2}{c^2}}$$

Where R' = shortened radius of particle

and R = original radius of particle

Since the mass is predicted to be inversely proportional to the radius then the following relationship holds:

$$\frac{R'}{R} = \frac{M}{M'}$$

Where M = the rest mass of the particle

(at rest relative to the observer)

And M' = the increased mass of the particle because of its motion relative to the observer.

The formula $$R' = R\sqrt{1 - \frac{v^2}{c^2}}$$

Can be rewritten as: $$\frac{R'}{R} = \sqrt{1 - \frac{v^2}{c^2}}$$

Now replace $\dfrac{R'}{R}$ with $\dfrac{M}{M'}$ and you get: $$\frac{M}{M'} = \sqrt{1 - \frac{v^2}{c^2}}$$

Solve for M' and you get: $$M' = \frac{M}{\sqrt{1 - \frac{v^2}{c^2}}}$$

If you solve this equation for a particle with mass approaching the speed of light (i.e. $v = c$) the result is this:

$$M' = \frac{M}{\sqrt{1 - \frac{c^2}{c^2}}} = \frac{M}{0} = \infty$$

This means that a particle approaching the speed of light would become infinitely massive, a clear impossibility.   But it is also clear that any mass moving at high speed versus a stationary observer will be seen to have an increase in its mass.

Mass Energy Equivalence

Using the Lorentz-Fitzgerald formula for determining mass increase Einstein derived what may be the most famous scientific formula of all time.   We'll see how he derived the mass-energy equivalence formula below.   You'll be surprised how easy it is.

$$M' = \frac{M}{\sqrt{1 - \frac{v^2}{c^2}}} \quad\quad \text{can be rewritten as:} \quad M' = M \cdot (1 - \frac{v^2}{c^2})^{-\frac{1}{2}}$$

Einstein next used a formula developed by Isaac Newton called the binomial theorem.

This theorem allows you to expand the algebraic expression of $(1 - \frac{v^2}{c^2})^{-\frac{1}{2}}$ into a infinite

series of terms with each successive term smaller than the previous one.   Applying the binomial theorem we get:

$$(1 - \frac{v^2}{c^2})^{-\frac{1}{2}} = 1 + \frac{\frac{1}{2}v^2}{c^2} + \frac{3v^4}{8c^4} + \text{.......}etc.$$

If we then substitute the expansion into the Lorentz formula we get:

$$M' = M\left(1 + \frac{\frac{1}{2}v^2}{c^2} + \frac{3v^4}{8c^4} + \ldots\ldots etc\right) = M + \frac{\frac{1}{2}Mv^2}{c^2} + \frac{3Mv^4}{8c^4} + \ldots etc$$

The third term of the expression and all others beyond it become so small for velocities below the speed of light that they can be ignored.

Next we notice that the term $\frac{1}{2}Mv^2$ in classical physics is the kinetic energy of a moving body. If we replace this term with the letter e for energy then we get:

$$M' = M + \frac{e}{c^2}$$

Or $$M' - M = \frac{e}{c^2}$$

$M' - M =$ the increase in mass of an object due to its motion.

Replace $M' - M$ with the term m to represent just the increase in mass due to motion. Then we get:

$$m = \frac{e}{c^2}$$

Or in its more familiar form of: $$e = mc^2$$

That's all there is to it.   Not that hard to do!

Einstein would later show that this formula that shows the equivalence of mass increase and energy applied to all masses, not just the special case of an increase in mass due to motion.

Summary

As strange as it may seem the constancy of the speed of light to all observers no matter what inertial reference frame they are in means that space and time must be variable.   However space and time taken *together* remain *invariant*.   In the next chapter we'll look at a neat way to see this space-time invariance which became instrumental in Einstein's development of general relativity.

There is no doubt the mathematics of special relativity works perfectly as a consequence of the speed of light being the same to all observers regardless of their relative motion.   Questions not often asked even in physics books are why the speed of light is constant and why space and time are separately variable?   What is the underlying physical cause for these phenomena?   As we shall see the answer to these questions will help to unravel the physics of the UFO enigma.

Minkowski Spacetime – Stepping Stone to General Relativity
Chapter 3 in Section II

Hermann Minkowski, who was one of Einstein's professors, developed an alternate way to describe special relativity. Minkowski realized that time could be viewed as a 4th dimension along with the three dimensions of space. This 4-D worldview is called spacetime, i.e. 3 space dimensions and one time dimension.

The idea behind the Minkowski spacetime diagram is that both space and time are not fundamental entities as was once believed before special relativity. Instead both space and time vary relative to each other but the combined quantity, spacetime, is invariant.

Because we live in three spatial dimensions it would be impossible to graphically show four dimensions. However we can use a simplified two dimensional world model consisting of one spatial dimension along with one dimension of time. This allows us to create a Minkowski spacetime diagram as shown below:

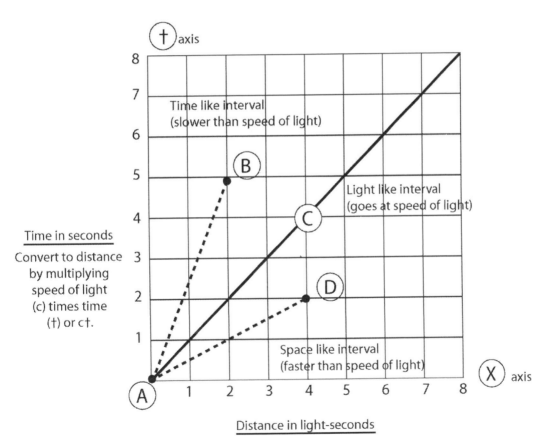

Time in seconds

Convert to distance by multiplying speed of light (c) times time (t) or ct.

Distance in light-seconds

One light-second equals distance light travels in one second or 186,000 miles.

In order to make *time* units on the **y** axis (labeled as **t** axis) compatible with the *distance* units on the **x** axis we can multiply the time units (seconds) by the speed of light **c**.   For here we will use **c=1** for convenience rather than 186,000 miles per second.

Time units converted to distance are then **c** times **t** or **ct** so in our case 1x1 second, 1x2 seconds, etc.   Since special relativity tells us that all observers measure the speed of light as **c** regardless of their motion then the speed of light becomes the ideal universal conversion factor to convert time units to space units.   Note that we used the distance light travels in one second on the **x** axis as 1 also, rather than 186,000 miles.   This way we are consistent and also end up with the same units on both axis which makes the chart easy to understand.

Imagine an event such as a flash of light occurring at point *A* on the spacetime diagram. After one second **(t=1)** on the vertical axis the flash of light can travel to the right only as far as one light-second (i.e. 186,000 miles) as shown on the **x** axis **(x=1)**.   At second 2 the light flash has expanded only as far as light-second 2 (2 times 186,000 miles) on the **x** axis.   And so on.   In the spacetime graph a flash of light, over time, can only go as far as the diagonal solid line shown on the spacetime diagram.   A photon traveling from point *A* to point *C* on this line is considered to have followed a *lightlike interval*.   Since we are not measuring normal spatial distance but rather distance in spacetime, we call this measurement a *spacetime interval* or just interval to distinguish it from ordinary distance.   The lightlike spacetime interval represents the path a particle (such as a photon or other massless particle) traveling at the speed of light follows in 4-D spacetime.

Next consider a material object traveling at constant speed starting at point *A* on the spacetime diagram and after 5 seconds **(t=5)** arriving at point *B* on the spacetime diagram.   A material object can only travel at less than the speed of light.   The spacetime interval between points *A* and *B* is called a *timelike interval* because of its two components (time **t** and distance **x**) the time component is dominant.   Note that **t=5** while distance of **x=2** is smaller.   All material objects follow timelike intervals through spacetime.

Finally consider an object that follows the spacetime interval going from point *A* to point *D*. It is clear such an object would have to travel faster than light.   Physicists speculate that particles known as tachyons could exist but cannot be detected since they always go faster than light.   Any object following a path in the spacetime diagram section below the lightlike interval line would be classified as traveling on a *spacelike interval* where distance (space) is the dominant component.

As we learned from the formulas governing special relativity a stationary observer will see time slow down and length contract for any object or person traveling at uniform velocity relative to themselves.   If an object could travel at the speed of light then time for that object would stop and length would contract to zero.   Minkowski realized that for his spacetime diagram to mimic the rules for special relativity he would have to make any lightlike spacetime interval equal to zero. This can be achieved by modifying the Pythagorean theorem to ensure lightlike spacetime intervals equal zero.

The lightlike interval from *A* to *C* in the spacetime diagram on the previous page can be determined by calculating the square root of the sum of the squares of the two legs of the right triangle formed by the hypoteneuse *A* to *C*.   We can see that this will be $c^2 t^2 + x^2 =$

$(AC)^2$. This is equal to: $4^2 + 4^2 = (AC)^2$. Solving for AC we get: $(AC)^2 = 32$ and then $AC = \sqrt{32}$. However in order to make AC equal to zero it is clear we need to subtract the square of one leg from the other (it doesn't matter which one is positive or negative). This modified Pythagorean theorem will then be as follows:

$$(AC)^2 = c^2 t^2 - x^2 = 4^2 - 4^2 = 0$$

Since our real world has three space dimensions and one time dimension the above formula for space dimensions **x**, **y** and **z** as well as time dimension **t** becomes:

Let **ds**=spacetime interval (**d** means difference in mathematics so **ds** here means the distance between two points in a spacetime interval).

The **x, y** and **z** axis will have "legs" for the **ds** spacetime interval of **dx, dy** and **dz.**

The time interval is **dt** which we convert to space units as **cdt.**

Then in three dimensions the spacetime interval formula becomes:

$$ds^2 = c^2 dt^2 - dx^2 - dy^2 - dz^2$$

(The above formula was devised to meet the requirements for calculating a lightlike spacetime interval but in fact it applies to spacelike or timelike intervals as well.)

This modified version of the Pythagorean theorem reproduces the math of special relativity! We can give another example to demonstrate this. Again for simplicity we will go back to our one spatial dimension and one time dimension model.

Imagine Fred and Alice on two different planets going by each other in opposite directions. Since Fred is on a planet he figures he is stationary and only moving through time. However he sees Alice not only moving through time but space as well as she whizzes by on her planet. Of course from Alice's point of view she thinks she is only going through time and it is Fred who is going through time and space.

Fred draws a Minkowski graph to represent Alice's motion through spacetime. This is shown below:

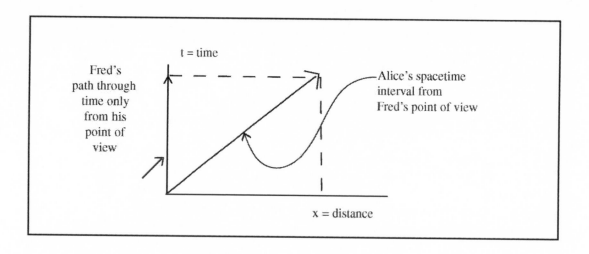

But Alice sees her own spacetime interval differently.   She figures she is only moving through time, that is she assumes she is only on the vertical **t** axis and simply measures the time on her clocks as we see in the Minkowski diagram below:

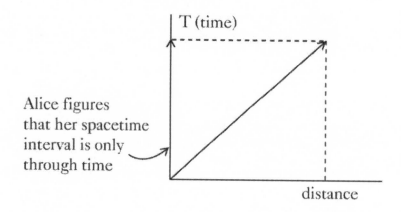

So Fred measures Alice's spacetime interval as:

$$ds^2 = c^2 dt^2 - dx^2$$

Simplify as: $s^2 = (ct)^2 - x^2$ (assume **s**, **t** and **x** are the measured "differences" of **ds, dt** and **dx**).

But distance **x** is the same as velocity times time or **vt**.   So now we have:

$$s^2 = (ct)^2 - (vt)^2$$

Which is what Fred measures as Alice's spacetime interval.

But Alice figures her spacetime interval as: $$s^2 = (ct)^2 - (0)^2$$

She figures she is only going through time so there is no space component in her spacetime interval calculation. Using her clocks she measures her time as (capital) **T**.

Now we know that all observers measuring the same spacetime interval will have the same

result. Since both Fred and Alice are measuring Alice's spacetime interval then their s^2 values will be the same. This means we can equate their spacetime interval formulas as follows:

$$(cT)^2 - (0)^2 = (ct)^2 - (vt)^2$$

Simplified: $$(cT)^2 = (ct)^2 - (vt)^2$$

Divide both sides by c^2.

$$\frac{c^2 T^2}{c^2} = \frac{c^2 t^2}{c^2} - \frac{v^2 t^2}{c^2}$$

Then: $$T^2 = t^2\left(1 - \frac{v^2}{c^2}\right)$$

Or: $$t^2 = \frac{T^2}{\left(1 - \dfrac{v^2}{c^2}\right)}$$

Then: $$t = \frac{T}{\sqrt{1 - \dfrac{v^2}{c^2}}}$$

This is, of course, the formula for special relativity time dilation!

It was Minkowski's marriage of space and time that helped Einstein formulate his theory of general relativity.   We'll see how Minkowski's concept of spacetime intervals is incorporated into Einstein's masterpiece in the next chapter.

The General Theory of Relativity
Chapter 4 in Section II

Curved Spacetime

Special relativity in 1905 changed our understanding of space and time but it was limited to observers in constant, non-accelerated motion with respect to each other. In addition, although the special theory brought into a single framework the concept of non-accelerated motion and Maxwell's laws of electromagnetism (i.e. the speed of light), it did not include accelerated motion and the force of gravity.

It is important to note here that in the early 20th century the known laws of physics included Newton's laws of motion, Einstein's special theory of relativity (which to some extent modified Newton's laws), Maxwell's theory of electromagnetism and Newton's laws explaining gravity. An understanding of nuclear forces and the associated science of quantum mechanics came at a later date.

In any case Einstein wondered how he could make Newton's law of gravity compatible with special relativity. Around 1907 it occurred to him that a person falling from a building does not experience any forces. A person in free fall in the earth's gravity field clearly accelerates as they approach the ground yet while they are falling they experience no forces at all. The force of gravity is exactly canceled by the acceleration of their fall. Furthermore the weight of the person does not change his rate of acceleration, a heavy person and a skinny person dropped from the same height both accelerate at the same rate. Galileo had noticed this effect using objects (rather than people!) centuries before.

Einstein quickly realized that in Newtonian mechanics acceleration and force are directly related in Newton's famous equation **F=ma**. Yet in a gravity field a person in free fall who is clearly accelerating does not feel *any forces*. Nor does the mass **m** of the falling object make a difference in the rate of acceleration. If ever there was a eureka moment this was it! In an instant Einstein realized that gravity is not caused by a force, at least not in the traditional sense.

Since a person in free fall experiences no forces that means that he meets the reference frame requirements of special relativity at any instant during his fall. Remember that special relativity only applies to observers not experiencing forces which produce acceleration of their reference frame. It is true that a person in free fall accelerates *between* reference frames but if his free fall is broken up into an infinite number of individual reference frames, then each of those frames can be thought of as obeying the requirements of special relativity.

Out of this thought process Einstein proposed his famous *equivalence principle*. This principle states that there is a "complete physical equivalence of a gravitational field and the corresponding acceleration of the reference frame. This assumption extends the principle of (special) relativity to the case of uniformly accelerated motion of the reference frame."

To put it differently a person standing on earth's surface experiencing a gravity "force" of one g is physically identical to another person in a rocket ship in deep space (far from any gravity field) and accelerating at 32 feet per second per second (or one g). The person in deep space will be pressed to the floor of the rocket ship cabin by the *inertial force* (defined earlier by Newton as resistance to change in motion) due to *acceleration*. Likewise the person "pressed" to earth's surface is feeling the resistance of the earth's surface due to his ***acceleration***.   Simply put this means the gravity force and inertial force are equivalent and both result from acceleration!

Einstein loved thought experiments and his chain of reasoning led to another gedanken (thought) experiment.   Imagine a physicist standing in an elevator whose cable has snapped causing the elevator to drop in free fall.   There is no experimental way the physicist can determine if he is in deep space and motionless or weightless due to dropping free fall in an elevator shaft that is embedded in a gravity field.   Under these conditions Newton's laws of motion apply.   If a light beam is flashed from a point on one side of the elevator it will hit the wall on the other side at the same distance up from the floor.   However if the elevator has glass walls an observer on the ground will see something interesting.   During the time the light beam crosses from one side of the elevator to the other side the elevator will have accelerated toward earth's surface some distance. In order for the light beam to hit the same spot on the opposite wall it will have to bend downwards to compensate for the downward movement of the elevator closer to the ground.   The ground observer will see the beam of light ***bend*** in the gravitational field of earth.   Thus Einstein reasoned that a gravity field will bend the path a light beam takes!

Now it is well known in the science of optics that light always follows the path that takes the least time, i.e. shortest distance.   In the flat space envisioned by Euclidean geometry the path of least time and shortest distance between two points is a straight line.   But here in a gravitational field it is apparent that the path of least time and shortest distance is a ***curved line***!   This is called a ***geodesic*** in the lingo of general relativity.

At the beginning of this chapter we saw that Einstein had initiated the above thought process in order to make Newton's law of gravity compatible with special relativity.   But special relativity is based on the concept of merging space and time in a combined entity that Minkowski called 4-D spacetime.   So it is this 4-D Minkowski spacetime rather than just space itself that is curved to produce the effect we call gravity.   Minkowski's 4-D spacetime concept had originally been dismissed by Einstein but now he realized its usefulness in general relativity.   However Minkowski's 4-D spacetime was designed for the flat (not curved) spacetime of special relativity. Einstein understood that he would have to incorporate the curvature of 4-D spacetime.

Besides explaining how curved spacetime is responsible for what we call gravity, Einstein also found he had to modify Newton's first law of motion.   That law says that a body will move along a straight line in space unless a force acts on it.   Einstein revised this to say that a body will move along the curve of shortest distance (the geodesic) in spacetime unless a force acts on it.

Now Einstein had the basic principles needed to modify Newtonian gravity to take into account special relativity.   However devising the math of a new theory of gravity was easier said than done.   It took almost ten difficult years to accomplish this task!

The Math of General Relativity

While searching for a way to express his theory of gravity mathematically Einstein's friend Marcel Grossmann introduced him to a type of geometry developed by the German mathematician Georg Riemann in the 1850's.   Riemann created the mathematical tools needed to deal with not only curved geometric spaces but also multidimensional spaces beyond the 3-D space of Euclidean geometry.   This was exactly the math Einstein needed to complete his theory.

The key to developing Einstein's general theory of relativity is a mathematical device Riemann had developed called a ***tensor***.   Tensors are nothing more than sets of rules for converting one math object into another.   Typically the rules are expressed in a matrix form such as a 3x3 matrix or 4x4 matrix etc.   At different locations in the rows and columns of the tensor matrix are the exact math rules for transforming one math object into another.   A tensor could be used, for example, to transform one 4-vector into another 4-vector.   Or a 4-vector into a scalar.   A scalar is just a number without direction such as temperature or someone's weight.   A vector in contrast has both direction and magnitude.   A weather map, for example, will show the direction of the wind at a given point and its magnitude.   When dealing in 3 dimensions a vector representing direction will be formally determined as the result of three perpendicular vectors along the x, y and z axis as shown in the diagram below.   The Pythagorean theorem extended to 3 dimensions is used to calculate the diagonal vector "**d**".   In the example below the perpendicular vectors have lengths **a, b** and **c**.   The diagonal is length **d**.   To calculate simply sum the squares of the perpendicular vectors and equate to the square of **d**.   Then solve for **d**.   So this gives

$$a^2 + b^2 + c^2 = d^2. \quad \text{Then} \quad d = \sqrt{a^2 + b^2 + c^2}$$

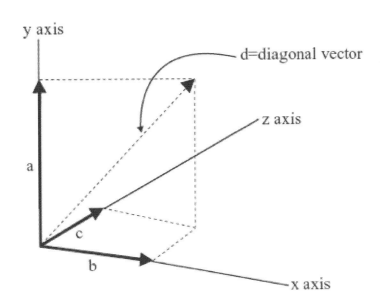

Although we cannot visualize a 4-D space it was easy for Riemann to extend the Pythagorean theorem to this situation simply by adding a fourth vector or term to the formula.   The Pythagorean theorem in 4-D would then be: $a^2 + b^2 + c^2 + d^2 = e^2$.

We have already used a tensor in the previous chapter on Minkowski spacetime.   As you will (hopefully!) recall we used a modified version of the Pythagorean theory to convert the 4-vector of spacetime (3 space dimensions and one time dimension) into a scalar called the spacetime interval. We modified the Pythagorean theorem by using opposite signs for the space components versus the time components in calculating the spacetime intervals.   This is an example of the special rules (or a tensor) that are used in this case to transform a spacetime 4-vector into a scalar spacetime interval. See, despite the fancy language, the concept of a tensor is not hard to understand.

The tensor that transforms a Minkowski 4-vector into a scalar spacetime interval is called the *metric tensor*. The metric describes the rules needed to modify the Pythagorean theorem to turn the spacetime 4-vector into a scalar. The tensor in this case is the 4x4 checkerboard that houses these rules.

Einstein knew that the key to the math of general relativity would require taking into account the various ways that spacetime could be curved. It was this curvature of spacetime that causes the effect we call gravity. The Minkowski 4-D spacetime is flat. However Riemann had already categorized the multitude of ways that space could be curved. He had investigated the curvature of spaces such as 2-D, 3-D, 4-D, etc. He found, for example, that he could describe the curvature at any point on a 2-D surface with just 3 numbers at each point. For a 4-D surface Riemann discovered that he needed 10 numbers at each point to describe the curvature. Generally the more dimensions of a particular surface the more numbers were needed to describe its curvature at each point.

For Einstein's purposes he needed the 4-D tensor developed by Riemann. This would equate to the 4-D of space and time but with time treated as another dimension of space. The 4-D tensor is illustrated below:

$$
\begin{pmatrix}
g_{11} & g_{12} & g_{13} & g_{14} \\
g_{21} & g_{22} & g_{23} & g_{24} \\
g_{31} & g_{32} & g_{33} & g_{34} \\
g_{41} & g_{42} & g_{43} & g_{44}
\end{pmatrix}
$$

This is Riemann's 4x4 metric tensor. The subscript indexes just indicate the position of each math operation. Only 10 numbers are needed to describe the curvature at each point of 4-D space. But note that there are 16 math components. However 6 of the 16 components are duplicates. For example $g_{21} = g_{12}$, $g_{31} = g_{13}$, etc. The net number of unique components is 10.

When mathematicians and scientists use tensors in equations they conserve space by using a symbol to represent a tensor. For example the 4x4 tensor above is represented by the letter "**g**" with two subscripts that we can label as "**i**" and "**j**". This would then look like: g_{ij}

The **i** represents the row number and the **j** represents the column number. The 4x4 tensor would then be described as: g_{ij} where **i** can be 1 to 4 and **j** can be 1 to 4.

Einstein knew that the curvature or warping of spacetime is what creates the effect we call gravity. But what causes spacetime to curve? It did not take long for Einstein to realize that it is the matter in a particular region that causes the curvature of spacetime. But as we learned in the chapter on special relativity, mass and energy are equivalent as shown in the formula

$$E = mc^2.$$ Therefore Einstein concluded that both mass and energy contribute to the warping of spacetime. But since matter is also likely to be moving in any region of space it was necessary to consider not just the matter itself but the motion of matter as well. In physics the motion of a mass is called its momentum which is simply its mass times its velocity $\left(m \times v\right)$. So the curvature of space in any region is seen to be the result of the energy in that region, mass and momentum. This then includes all forms of energy as well as mass.

After much effort Einstein devised an equation that describes how matter and energy in a particular region of spacetime cause the curvature of spacetime in that same region. Like Maxwell's equations of electromagnetism the equations of general relativity operate at a local level, that is at each point in spacetime. This is quite different from Newton's law of gravity that assumed an instantaneous gravitational force between distant masses. In general relativity by contrast the gravitational effect cannot exceed the speed of light as required by special relativity. The math of general relativity determines the overall curvature of a region of space as a result of the cumulative effect of a vast number of Riemann points in space each of which has some degree of curvature.

The starting formula for general relativity consists of equating two different tensors that relate the mass and energy in a region of spacetime to the (curved) structure of spacetime at the same location. The Einstein curvature tensor is set equal to the stress-energy tensor in what is known as the **Einstein field equations (EFE)**. The EFE is shown below:

$$G_{ij} = 8\pi G T_{ij}$$

G_{ij} is Einstein's curvature tensor that calculates the curvature of spacetime at each point.

$8\pi G$ this entire term is a constant. The G here is Newton's gravity constant.

T_{ij} This is the stress-energy tensor which determines the energy and momentum in a region of spacetime.

The expanded form of the EFE reveals all the components of the tensors as shown below:

$$\begin{pmatrix} G_{11} & G_{12} & G_{13} & G_{14} \\ G_{21} & G_{22} & G_{23} & G_{24} \\ G_{31} & G_{32} & G_{33} & G_{34} \\ G_{41} & G_{42} & G_{43} & G_{44} \end{pmatrix} = 8\pi G \begin{pmatrix} T_{11} & T_{12} & T_{13} & T_{14} \\ T_{21} & T_{22} & T_{23} & T_{24} \\ T_{31} & T_{32} & T_{33} & T_{34} \\ T_{41} & T_{42} & T_{43} & T_{44} \end{pmatrix}$$

There are 10 independent terms in each tensor out of the 16 terms in the 4x4 tensors as explained before. This results in 10 independent equations that must be solved simultaneously. And you need a separate complete equation with 10 terms in each tensor for each point in the spacetime region you are studying! But not to worry, I will just illustrate the 10 independent equations below for just one point in spacetime. We won't try to use real data and attempt an actual solution since in reality the EFE is very difficult to solve! Here's the 10 independent generic equations for one point in spacetime:

1) $G_{11} = 8\pi G T_{11}$

2) $G_{12} = 8\pi G T_{12}$

3) $G_{13} = 8\pi G T_{13}$

4) $G_{14} = 8\pi G T_{14}$

5) $G_{22} = 8\pi G T_{22}$

6) $G_{23} = 8\pi G T_{23}$

7) $G_{24} = 8\pi G T_{24}$

8) $G_{33} = 8\pi G T_{33}$

9) $G_{34} = 8\pi G T_{34}$

10) $G_{44} = 8\pi G T_{44}$

The EFE are a bit more complicated than I've shown. The Einstein curvature tensor is actually comprised of two other tensors which would each have to be expanded into the 10 independent components. The full Einstein curvature tensor is this:

$$G_{ij} = R_{ij} - \frac{1}{2} R g_{ij}$$

Where G_{ij} = Einstein curvature tensor

R_{ij} = Ricci tensor

R = scalar curvature

g_{ij} = metric tensor

In addition Einstein realized his original equations could result in the math indicating the universe was expanding or contracting. Since he assumed the universe was static he added the term Λg_{ij} to ensure a static universe. So then the final equation for the EFE would be as follows:

$$R_{ij} - \frac{1}{2} R g_{ij} = 8\pi G T_{ij} - \Lambda g_{ij}$$

The term "$-\Lambda g_{ij}$" was included on the "stress-energy" side of the EFE to create a static universe. The component "Λ" of the previous term is called the *cosmological constant*.

Some ten years after the EFE was published it was discovered by Edwin Hubble that the universe was not static but is in fact expanding. Einstein then called his inclusion of the cosmological constant term "the biggest blunder I ever made". However in more recent years it has been discovered that there is a need for the cosmological term. When this term is included in the EFE you can derive a constant called the **vacuum energy** which is important in theories of the universe. The vacuum energy and cosmological constant terms are often used interchangeably.

Connections

After solving the Einstein field equations to determine the Einstein curvature tensor we are ready for the next step in the general relativity process. (Yup, there's another step...but we're almost there!) In flat space a vector tangent to the flat space representing a particular direction in that space will stay parallel to its original position if moved parallel to itself. So if you imagine a vector (arrow) on the bottom of a flat 8 by 12 sheet of paper (a 2-D space analogy) pointing toward the top of the sheet, it will remain pointing toward the top of the sheet if its moved anywhere on the sheet but kept tangent to the sheet and parallel to its initial direction. This is shown in the illustration below:

Parallel Transport of Vector in Flat Spacetime

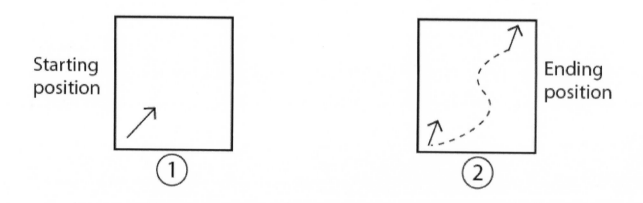

In curved spacetime, however, parallel transport will usually result in a 4-D spacetime vector pointing in a different direction. We cannot visualize a 4-D spacetime so like the previous example we'll use a curved 2-D space to demonstrate parallel transport in curved space.

The surface of a sphere is a 2-D space which is curved and is easy to visualize. Imagine a tangent vector pointing south at the north pole of the sphere. First move it directly south down a meridian till it hits the equator. Next try a different route for the vector by moving it from the north pole down a different meridian that is perpendicular to the first one. Once you hit the equator using this route then move the vector to the point on the equator where the vector was first moved.   You can see in the illustration below that the vector ends up pointing in two different directions depending on which route was chosen.

Parallel Transport in Curved Spacetime

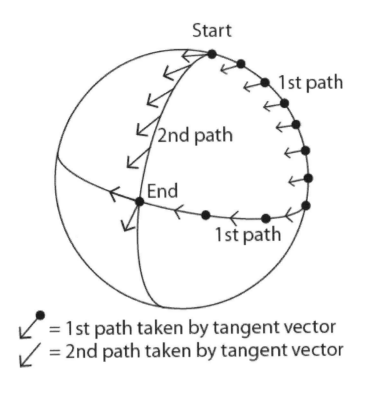

Note that depending on the path taken the vector ends up pointing in two different directions at its final location.   This is unlike the situation of parallel transport in flat spacetime where the vector always ends up parallel to its original direction.

In curved spacetime a parallel transported tangent vector moved along various paths from a given starting point to a given end point will in general end up pointing in just two particular but different directions.   The difference between these two end point vector directions can be measured by a math object called a ***connection***.   For any region of spacetime being investigated there will be a complete set of these connections.

The final step of the general relativity process converts these connections into terms that are components of a metric.   These components are then used to determine all possible spacetime intervals or geodesics between all pairs of points in the spacetime region under study.

Einstein's general theory of relativity ultimately describes a two way "dance" between spacetime and matter-energy.   Spacetime determines how matter moves and matter (and energy) forces spacetime to curve in a particular way.

An easy way to see how the curvature of spacetime creates the effect we call gravity is to again use the 2-D "spacetime" earthlike sphere we used in the parallel transport example.   Imagine two people named Fred and Alice each located at different points on the equator, say one quarter of the distance around the equator from each other.   To start with both Alice and Fred are facing due north and are therefore parallel to each other at the equator although separated by one fourth of the equator's circumference.   If both Alice and Fred begin moving north at the same speed they will notice that they are gradually getting closer to each other until they finally bump into one another at the north pole.   Unless they understood the curved geometry of the earthlike 'space' they would assume some force called gravity pulled them together.   In reality it was the curvature of the 2-D sphere 'space' that caused the effect we call gravity.

Conclusion

General relativity ranks as one of the greatest accomplishments in the history of science.   It has been an essential tool in studying the universe at large and was instrumental in predicting and studying black holes and other unusual phenomena.   Einstein will be forever linked with this extraordinary achievement.

Despite the great success of general relativity in explaining gravity there still remain some unanswered questions about Einstein's masterpiece.   The important question for this book is Einstein's postulate equating gravitational mass and inertial mass.   Bear in mind that a postulate is an assumption, not a proven fact.   Although all experiments indicate that gravitational mass and inertial mass are equivalent there is no fundamental physics theory that explains why this is so. They could have been different but for some reason mother nature has made them the same.

The equivalence of gravitational and inertial mass is, I believe, important to the technology behind the UFO phenomenon.   Later in the book we'll look at a theory that incorporates the latest ideas in contemporary physics that may explain the equivalence principle as well as its role in how UFOs operate.

In the next chapter we'll take a look at the subatomic building blocks of our world.   This background information will be essential for understanding the branch of physics known as quantum mechanics.

The Building Blocks of Our World
Chapter 5 in Section II

The Basic Constituents of Nature

Although our world is tremendously complex and varied the building blocks that make up our world are relatively simple.   For example, the matter or elements that comprises everything we are familiar with such as iron, copper, carbon, oxygen, hydrogen etc. is constructed from only three basic particles at the atomic level.   These are protons, electrons and neutrons.   The atoms themselves have been traditionally thought of as consisting of a nucleus that contains protons and usually neutrons with electrons orbiting the nucleus.   The protons, each with a positive charge of +1 are normally balanced in the atom by an equal number of electrons which each have a negative charge of -1.   Neutrons, as the name implies, have no electric charge.   Protons and neutrons are roughly 2,000 times more massive than an electron.   Nevertheless all the subatomic particles are incredibly tiny.

The known forces that govern the interaction of all matter particles number only four in total. Two of these forces are familiar to us as gravity and electromagnetism.   The two other fundamental forces, called the strong nuclear force and the weak nuclear force, operate only at the subatomic level within the nucleus of atoms.   A fifth force, not yet confirmed, is known as the Higgs field.   It is believed to play a role in giving mass to the matter and force particles.   More on this fifth force later.

The Matter Particles of Nature

We'll first examine the matter particles that are the constituents of all the things that are familiar to us such as wood, metal, glass and the very air we breathe.   Of the three basic particles; electrons, protons, and neutrons that are the building blocks of matter, only electrons are fundamental.   After years of research scientists have discovered that protons and neutrons are themselves made up of more fundamental entities called *quarks*.   This means that all the matter we are familiar with is composed of only two fundamental building blocks: electrons and quarks.

Scientists have also discovered that there are six types of quarks which are humorously labeled as the up, down, charm, strange, top and bottom quarks.   Physicists were surprised to find that quarks have fractional charge.   The up, charm and top quarks have electric charge of +2/3 each while the down, strange and bottom quarks have electric charge of -1/3 each.   Since the proton and neutron have +1 and zero electric charge that meant only a certain combination of quarks could account for their whole number charge.   It was found that a proton, for example, consists of two up quarks and one down quark.   This translates to quark charges of +2/3 and + 2/3 less the -1/3 for a net charge of +1.

It has also been determined that the six quarks can be classified into three groups or families of two each based on increasing mass and other characteristics.   The first and lightest quark family is the up and down quark, the second is the charm and strange quark and the last and most massive is the top and bottom quark.   Scientists have found that only the first family of quarks are stable. The other two families quickly decay into their stable, lowest mass cousins.   Needless to say normal matter is made up of the first family of quarks.

Scientists later discovered an additional type of matter particle called the neutrino which is associated with the electron in certain types of particle interactions.   Further research revealed what appeared to be two more massive types of electrons, one heavier than the other, which were labeled the muon and tau.   Each had the same charge of -1 as the ordinary electron and were identical in all other respects except for mass.   Perhaps not surprisingly there were two additional types of neutrinos discovered, one associated with the muon, called the muon neutrino and the other associated with the tau and called the tau neutrino.   All the neutrinos have neutral charge and are believed to have only a tiny mass.   Like the quarks the three types of electrons and their associated neutrinos are arranged in three families based on the increasing mass of the electron types as well as certain particle characteristics and interactions.

Scientists now summarize all matter particles into two groups, quarks and **leptons**.   There are a total of six quark types in the quark category.   Likewise there are six leptons in the lepton category which includes the three types of electrons and three types of neutrinos.   In turn all matter particles, both leptons and quarks, are classified in a single category called fermions named after the famous Italian physicist Enrico Fermi.

So is that all there is to the matter of the universe?   Well not quite.   There is at least one other important characteristic of these fundamental matter particles that needs to be mentioned. During the middle part of the last century physicists had noticed that subatomic particles subjected to magnetic fields behaved in an odd manner.   It was concluded that the fundamental matter particles have a intrinsic spin that is unlike the kind of spin associated with classical physics (i.e. Newtonian mechanics and Einstein's relativity theories).   This intrinsic spin of the fundamental matter particles is part of the newer, non-classical branch of physics we call quantum mechanics.

A classical object such as a children's top will spin until friction slows it down and it stops spinning after toppling over.   However quantum spin is very peculiar in that the fundamental particles never stop spinning.   As if this were not weird enough scientists also noticed that fermions (matter particles) have to rotate twice through 360 degrees (in magnetic fields) to get back to their original starting point!   So this means fermions have to rotate 720 degrres in order to complete "one" complete rotation.   In quantum theory the intrinsic spin of fermions is labeled as ½ (ie. it takes two rotations to complete one complete turn: 1/2+1/2=1).   This quantum spin of 1/2 is considered to be a fundamental characteristic of fermions just like electric charge and mass.

There is one additional characteristic of matter particles that also needs to be mentioned.   All matter particles come in two versions, the particle and its anti-particle.   Each anti-particle is identical to the corresponding particles described above except that the electric charge is reversed. For example the antiparticle of the electron, called the positron, has electric charge of +1.   The anti-particle of the neutron has opposite spin.

With the exception of one more interesting characteristic of quarks that we will cover shortly, the above covers essentially all that is known about the fundamental matter particles in the universe. Next we'll take a look at the force particles that are the vital "glue" that holds our world together.

The Force Particles of Nature

All the particles in nature interact with one another by the four fundamental forces of electromagnetism, gravity, the weak nuclear force and the strong nuclear force.   As noted before a fifth force or interaction known as the Higgs field, which is believed to be involved in giving mass to many of the fundamental particles, may also exist.   All the forces manifest themselves in the form of so called exchange particles.   The electromagnetic force, for example, is carried by the zero mass photon.   Like the fermions the force particles also have quantum spin.   The photon has been measured to have quantum spin of 1.   The electromagnetic force operates between electrically charged particles.   An example would be when you put your hand on a desk.   Like electric charges are repulsive and opposite electric charges attract.   As your hand touches the desk the outer electrons in the atoms of the desk's surface repulse the outer electrons of your hand and prevent your hand from penetrating the desk.

An important aspect of the exchange particles needs to be introduced at this point. The photons that actually carry the electromagnetic force are *virtual* photons as distinguished from real photons. Virtual photons are photons that come into existence only momentarily. If two electrons approach each other closely then some of the vast sea of virtual photons between them will be "borrowed" to carry the repulsive force between the electrons. This same process works for all the exchange forces. Fermions, the matter particles, can also be manifested as virtual particles. An electron, for example, can quickly appear and disappear out of the vacuum of empty space. However if enough energy is added to a virtual particle it can become a real particle.

Virtual particles arise from the principles of quantum mechanics, the non-classical branch of modern physics. Werner Heisenberg, a German physicist who was one of the founders of quantum theory in the 20th century, explained virtual particles in terms of his uncertainty principle. Quantum mechanics envisions the vacuum of space as consisting of a froth of virtual particles that come into existence by momentarily borrowing energy from the quantum vacuum of space in a balanced manner. For example a virtual electron can come into existence if it is accompanied by a virtual positron and if both particles return to the vacuum almost immediately. A photon, which has no electric charge of its own although it can carry charge between charged particles, is its own anti-particle. This means it can arise by itself from the quantum vacuum as long as it gives back its borrowed energy quickly and returns to the vacuum. We'll look at the Heisenberg uncertainty principle again in Section II chapter 6 on quantum mechanics.

The gravity force is represented by an exchange particle called the *graviton* in quantum theory. However Einstein's general theory of relativity in the previous chapter explains gravity in terms of the geometry of spacetime. As yet there is no quantum theory of gravity but recent theories do predict the existence of the spin 2 graviton. The pursuit of a quantum theory of gravity remains as a kind of Holy Grail in modern physics.

The weak nuclear force is manifested by three exchange particles designated as the W^+, W^- and Z^0. The weak nuclear force is responsible for radioactive decay in atoms of elements such as uranium.

Quarks are known to be constituents of protons and neutrons as well as other more exotic particles. But what holds the quarks of the same electric charge together since ordinary particles of the same charge should be repulsive? Furthermore what holds the positively charged protons together in the nucleus of an atom?

After a great deal of experiment and research physicists identified a type of particle called the gluon which is responsible for the strong nuclear force that holds the quarks together as well as the protons and neutrons in the nucleus of an atom. The electromagnetic and gravity force get weaker with distance based on the inverse square law. The strong force, however, gets stronger with distance.

Scientists have identified eight types of gluons. The gluons carry a type of charge called color charge. There are three color charges which are labeled for convenience as red, blue and green. Just as the photon exchanges electric charge between electrically charged particles such as electrons and protons, gluons exchange color charge between color charged nucleons like the proton, neutron and their constituent quarks.

As a result of the discovery of gluons, physicists realized that quarks contain color charge of red, blue or green. However it was also realized that nucleons are color neutral which means the color charges of the constituent quarks of a nucleon must sum up to color neutral. This is why the three primary colors of red, blue and green were chosen since in the art world these three colors together result in the color white, a neutral color.   Since both the neutron and proton are comprised of three quarks each then each quark must be a different color to ensure these nucleons are color neutral.

Some composite particles are made of just two quarks.   These are called mesons and they are involved in transmitting the strong interaction within the nucleus.   Since mesons are color neutral it was found they consisted of a quark and an anti-quark, for example a red quark and an anti-red quark.

All the non-gravity force particles have quantum spin of one.   These force carriers for electromagnetism, the weak nuclear force and strong nuclear force are called **bosons** in honor of Satyendra Bose, an Indian physicist who did pioneering work on quantum particles with integer spin.   The graviton with predicted spin of 2 is also considered a boson.

This concludes our review of the particles involved in transmitting the forces of nature.   Only the postulated Higgs field/particle has not been covered but this will be examined in a later chapter.

A Summary of the Matter and Force Particles

The charts below outline the known force and matter particles.

Force Particles (Bosons)

γ photon	g gluon	Z^0 weak−force	$W^\pm$ weak−force

Note: 1) All these force particles have spin of 1.   2) The W particle can have weak charge of +1 or -1.

Three Generations of Matter Particles (Fermions)

I	**II**	**III**
u up	c charm	t top
d down	s strange	b bottom
ν_e electron neutrino	ν_μ muon neutrino	ν_τ tau neutrino
e electron	μ muon	τ tau

Note: 1) The **u, c** and **t** quarks all have electric charge of **+2/3**.   The **d, s** and **b** quarks have electric charge of **-1/3**.

2) All the matter particles above have spin of **1/2**.

General Note: All the matter and force particles have anti-particles as well.   The charge is reversed in the antiparticle where it applies (photons are their own anti-particle).   The mass of the particle and anti-particle is the same.

Quantum Mechanics
Chapter 6 in Section II

The Discovery of Quanta

Quantum mechanics had its origins in the physicist Max Planck's discovery in 1900 that energy is not continuous but comes in bits or chunks called *quanta*.   Unfortunately this did not fit into the classical physics view of nature that had been developed over many centuries.   The idea of energy existing as quanta instead of being continuous came as a shock to the scientific community of the time.

During the 19th century scientists had noted in experiments that when an object is heated to a temperature above its surrounding temperature it begins to radiate energy in the form of heat or infrared energy.   As the temperature of the object is raised even higher the object will radiate electromagnetic energy in both the infrared portion of the spectrum and also in the visible portion and eventually in the ultraviolet range as well.

Typically these experiments were done using a ceramic container with a small hole in one end. Light beams aimed into the hole would be trapped as they bounced around inside the cavity and could not escape.   But when the container is heated the radiation produced inside bounces around but eventually escapes through the hole.   This device was called a "black body" because it would absorb *all* wavelengths of light when cool and then emit *all* wavelengths when heated.

The key observation about black body radiation is that its characteristics depend only on temperature.   Black body radiation when graphed looks like a normal curve with its peak light frequency determined from the temperature of the black body.   The higher the temperature the further to the right the curve (range of frequencies) will move and the peak frequency will move a commensurate distance to the right as well.   A typical black body graph looks like this:

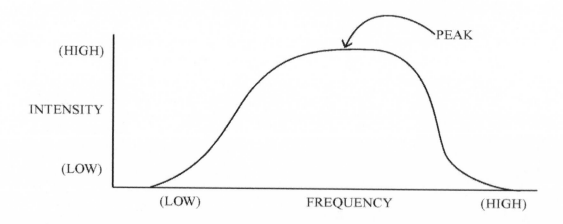

Typical Black Body Electromagnetic Radiation Distribution

The problem facing physicists at the end of the 19th century was that they could not predict the shape of the black body curve using Maxwell's wave equations of electromagnetism.   Classical electromagnetic theory predicted that the amount (intensity) of radiation should be proportional to frequency.   This just means that the higher the frequency then the greater the amount (or intensity) of the radiation emitted at that frequency.   According to classical theory the graph of radiation distribution should look like this:

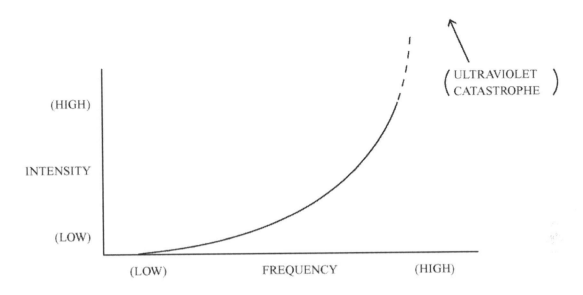

Classical Theory Predicted Black Body EM Radiation Distribution

Higher frequency is equivalent to shorter wavelength.   But classical theory also did not put any limit on how short wavelength could be.   This meant that a heated black body should emit virtually all its radiation at extremely short wavelengths and there would be no limit as to the amount of radiation emitted at these higher frequencies .   The result would be that a heated black body should radiate nearly all its energy almost immediately at the higher frequencies.   This came to be known as the "ultraviolet catastrophe" because any hot body, including stars, should emit all their energy virtually instantaneously in a blast of ultraviolet light!   But actual observation of hot objects always resulted in the typical bell shaped normal curve with radiation intensity falling off at the higher frequencies.   Thank goodness!   Nevertheless the disconnect between theory and observation left physicists puzzled.

The German physicist Max Planck finally solved the puzzle of black body radiation in 1900. Planck theorized that energy was not continuous like a wave but instead came in small units he called *quanta*.   The quanta for light came to be called *photons* just as the gravitational quanta are called gravitons.   He further theorized that the shorter the wavelength the larger the energy per quanta.   Planck's theory immediately explained the fall off in radiation at the shorter wavelengths. It was clear that at the shorter wavelengths, or higher frequency, the quanta required so much energy per quanta to be radiated that relatively few are emitted.

As I mentioned at the beginning of the book it is important to have a basic familiarity with the math of physics.   Math is the language of science and it's really not that hard once you get used to it.   Using the math makes it easier to see how physics evolved as new ideas were introduced.

With this in mind the actual formula that Planck derived is as follows:

$$E = hf$$

E = energy
h = Planck's constant
f = frequency of electromagnetic energy (full wave cycles per second)

Since Planck's constant h is very small this had prevented scientists from discovering earlier that energy flow is not continuous but instead broken up into smaller pieces (quanta) that cannot be subdivided further.   Like the atoms of matter the carriers of energy, such as photons, reach a limiting finite size determined by Planck's constant.   Now it was clear that both matter *and* energy come in bits or pieces.   See Appendix B for common length and energy scales in physics.

Although Planck did not realize it, his solution to black body radiation at the turn of the twentieth century was a major watershed.   It marked the end of classical physics and the beginning of modern physics.

Quantum Mechanics, the Atom and Neils Bohr

In the early years of the 20th century the English physicist Ernst Rutherford had devised a model that was thought to explain the structure of atoms.   Rutherford's model postulated that atoms consisted of a massive nucleus composed of protons that was surrounded by electrons which orbited the nucleus at various distances.   This appeared to explain much of the experimental data accumulated over the previous decades.

However as physicists looked deeper into Rutherford's model of the atom they began to see serious problems.   For example it was known from Maxwell's theory of electromagnetism that electrons give off light as they accelerate.   Therefore the electrons orbiting the nucleus should give off light waves since any object following a curved path at constant speed is experiencing a form of acceleration.   It was also thought that electrons orbiting the nucleus at different speeds should give off light of different frequencies.   An electron orbiting at higher speed should give off higher frequency light waves than one orbiting at lower speeds.   The basic idea is illustrated here:

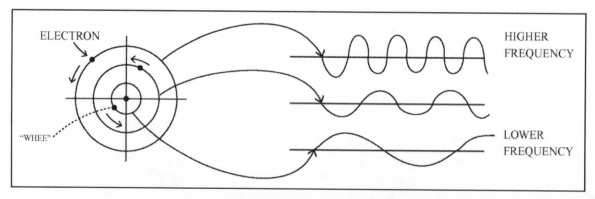

Since scientists knew that many atoms had numerous electrons in orbit around the nucleus then each type of atom should produce a unique spectrum after passing light emitted from th at atom through a prism.   This is illustrated below:

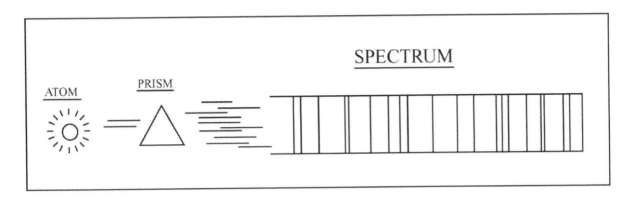

Furthermore, according to classical theory, as the electrons lost energy after emitting light waves they should begin falling inward closer to the nucleus and eventually all atoms should collapse completely!   Obviously this doesn't happen and atoms observed at room temperature are completely stable and do not emit light waves at all.   Only if atoms, such as a sample of hydrogen gas, are heated significantly or cooled off from a higher temperature do they emit or absorb light waves that show a unique spectrum for each type of atom.   But atoms at room temperature, despite having electrons accelerating around the nucleus, do not emit light waves as classical theory would predict.   Clearly a new explanation was needed to explain the behavior of atoms.

The Danish physicist Niels Bohr after reviewing various spectroscopic data proposed in 1913 that you cannot use classical physics alone to describe the behavior of electrons in an atom. Instead you must resort to the concept of quanta developed by Planck.

Bohr suggested that electrons occupy various orbits at certain fixed distances from the nucleus corresponding to the electron shells postulated by chemists.   He proposed that an electron will absorb electromagnetic energy when it moves outward from one orbit to the next or emit electromagnetic energy if it moves inward to an orbit closer to the nucleus.   Furthermore Bohr realized that the emission or absorption of specific wavelengths of light was related to quantum theory.   For a particular atom, only the absorption of energy quanta (wavelengths) of a specific amount would cause an electron to jump into the next orbit.   And vice versa.   Thus electrons could not occupy any orbit, but only orbits at certain fixed distances from the nucleus.   Electrons in a given orbit, however, were stable for some reason and did not emit light despite the assumption that they were orbiting the nucleus and thus experiencing acceleration.   Only the movement of electrons *between* orbits caused them to emit or absorb energy.

After a great deal of effort and incorporating Planck's quantum formula of $E = hf$, Bohr was able to explain mathematically much of the atom's behavior.   Some of the key achievements of Bohr's research are listed here:

1) He came up with a formula for determining the energy of electrons in various orbits.   An important discovery from this work was that the energy of electrons in outer orbits varied very little from orbit to orbit.   This turned out to be very useful in the further development of quantum mechanics.

The electrons energy, when graphed, resembled a step function like this:

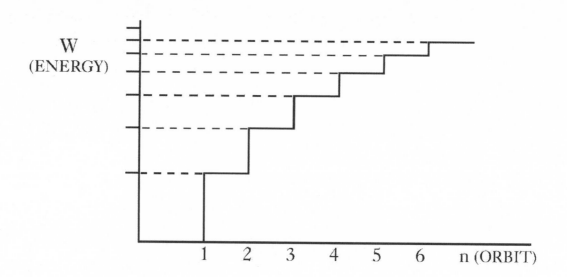

2) Using the difference in energy levels between electron orbits he was able to calculate the frequency of light given off as electrons transitioned from an outer (higher energy) orbit to an inner (lower energy) orbit.   The equation he devised to calculate frequency is this:

$$f = \frac{W_n}{h} - \frac{W_m}{h}$$

Where: $\quad W_n = \dfrac{-Rhc}{n^2}$ $\qquad$ **(Use the corresponding formula for W_m too.)**

$f =$ frequency

$R =$ Rydberg's constant **(determined from previous experiments)**

$h =$ Planck's constant

$W_n =$ Energy of electron in orbit n

n or m = Orbit number of electron where $n > m$.   When an electron transitions from an outer orbit *n* to a inner orbit *m* it gives off light of frequency f in the formula above. *n* and *m* could be (respectively) the number pairs of say 4 and 2, or 3 and 1, or 5 and 4 etc.

For outer orbits *only* he was also able to calculate the *intensity* of the spectra using another formula we'll look at shortly.

3) Bohr devised a formula that gave the angular momentum M of electrons in orbit around the nucleus as shown here:

$$M = \frac{h}{2\pi} n$$

He realized that this was a fundamental formula describing the subatomic world in terms of quantum theory with the use of Planck's constant h and orbit number n.   He decided to call this important general quantum law the **quantum condition**.

4) However when physicist Arnold Sommerfeld saw Bohr's formula he immediately realized that the formula for angular momentum only covers circular and elliptical motion but nothing else. Sommerfeld suggested they devise a more general description of the electron's momentum covering other ways quanta could move.   Using a basic formula for harmonic oscillation they devised a new, more comprehensive formula for the quantum condition.   This is that formula:

$$\int pdq = nh \qquad \text{(n=1, 2, 3, 4 etc)}$$

q = **position of electron**

dq = **change of position of electron**

p = **momentum of electron (mass times velocity)**

h = **Planck's constant**

n = **is any integer such as 1, 2, 3 etc.**

$\int$ = integration symbol, just means calculating the area under the curve associated with the formula for the quantum condition.   This area then represents the total momentum of the electron as it circles the nucleus in a particular orbit n.

The efforts of Bohr and others had now explained much of the mystery of subatomic behavior such as the stability of the atom, the radius of the atom, the frequency of an atom's spectrum and the previously puzzling discrete energy condition of atoms where energy varied according to the integer n was now understood to be the result of varying orbits of the electrons.

Despite the progress a major problem remained unaddressed.   Although Bohr could now predict the frequency of atomic spectra he was unable to come up with a formula that gave the intensity of the light at each frequency except when n is large.   When the electrons were in large n (outer) orbits a formula known as the *Fourier equation* did give the intensity for electron transitions.   Bohr discovered that when n is large the frequencies of the atomic spectra are integral multiples of a base frequency.   This allowed him to use the Fourier equation to determine the amplitudes of the range of atomic spectra but only when n is large.   The amplitude when squared gives the intensity for a particular spectra.

The Fourier Equation

So how does the Fourier equation determine the intensities of light spectra from atoms, at least when n is large?   In classical theory a complex light wave given off by a heated atom should break down into a set of frequencies that are *integral* multiples of some fundamental frequency. For example if the fundamental frequency is 2 cycles per second then integral multiples of the fundamental frequency would be 4 cycles per second, 6 cycles per second, 8, 10, 12, etc.

The Fourier equation was developed by the French mathematician Joseph Fourier in the early 19th century and should, according to classical physics, explain both the frequencies and amplitudes of the atoms spectra.   Fourier math was the only known way to understand complex iterative (repeating) waves of any type.   The idea was that electrons in a particular single orbit around the atomic nucleus are experiencing acceleration which causes them to emit complex light waves which should break down via Fourier into evenly spaced frequencies that can be observed in that atom's spectra.   In actual practice usually a sample of gas, such as hydrogen, is heated and the light given off is directed through a glass prism to reveal its spectrum.   The so called Fourier series equation should mathematically describe the spectrum frequencies and amplitudes as shown in the example below:

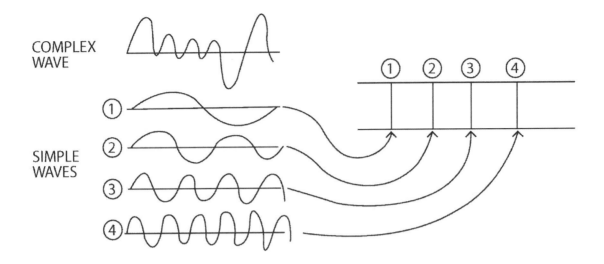

The complex wave would be the sum of the simple waves.   Mathematically the Fourier equation looks like this:

$$q = \sum_{\tau} Q(n,\tau)e^{i2\pi f(n,\tau)t}$$

q = position or amplitude of electron as reflected in the complex iterative lightwave emitted by the atom.   So if the complex wave looks like this:

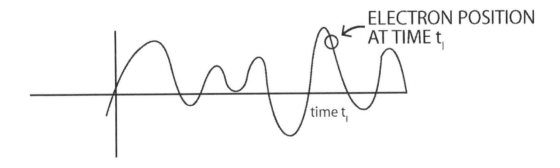

Then the amplitude at time t is this dot above the x axis.

$\sum$=sum of. This means the position "*q*" above is determined by summing all the simple waves at a given time "*t*". Say only two simple waves are being summed, one with a q value of -3, the other with a q value of +5, and both occurring at time "*t*". The sum would be +2 at time *t*. This means at time *t* the electron is located at q=+2 above the x axis (as reflected in the complex light wave given off by the atom which should mimic where the electron is located inside the atom).

$Q(n,\tau) =$ this term is the amplitude of the τth simple wave (i.e. the 1st or 2nd or 3rd etc simple wave in the series) that is part of the complicated light wave emittedby the electron which is in a particular (single) orbit *n*.

$f(n,\tau) =$ the frequency of the τth simple wave that is part of the complicated light wave emitted by the electron which is in a particular (single) orbit n.

$e^{i2\pi f(n,\tau)t} =$ this expression describes the general shape of the τth simple sinusoidal wave. The letter "*i*" is a mathematical constant which equals the square root of -1. π or "pi" is a constant which equals the circumference of a circle divided by its diameter. $f(n,\tau)$ is the frequency of the wave. The letter "*t*" is the variable time.

The classical Fourier equation was the only known method to determine the amplitude of a complex iterative wave and to break it down into its constituent simple waves and their associated amplitudes and frequencies. But it only worked for large *n* when trying to predict the frequency and amplitude of atomic spectra. For smaller *n*, frequency could only be determined by using Bohr's formula. Unfortunately there was no way to determine the amplitude of atomic spectra when *n* was small.

A frustrated Bohr now turned to his younger colleagues at the Copenhagen Institute to see if they could come up with the key to understanding the quantum world. He posted a message on his laboratory door challenging them to find a method for explaining both the frequency distribution and amplitude of all atomic spectra whether n was large or small.

A young German doctoral student by the name of Werner Heisenberg saw Bohr's message and decided to take up the challenge. Suffering from a severe bout of hay fever, Heisenberg decided to retreat to a pollen free island in the North Sea to work on the problem.

Heisenberg Solves Quantum Mechanics

The challenge facing Heisenberg was finding spectral frequency *and* intensity using quantum physics whether n was large or small. Heisenberg realized he had to incorporate the different behavior of electrons in quantum theory versus their theorized behavior in classical physics.

In classical theory an electron was thought to stay in a particular orbit but its complex motion created a single complex light wave that in a spectrum broke down into a multitude of single frequencies that are integer multiples of a base frequency. There would be varying intensities for the different frequency lines of the spectrum. Fourier math could be incorporated in solving for spectral intensity when n was large because when the electron is in a large n (outer) orbit it only has to transition a tiny distance between orbits to emit or absorb photons. It turns out these transitions are not only tiny but evenly spaced too so that it appears as if the electron is in a single orbit with complex motion whose light emissions can be broken down using Fourier math. But the merging of classical physics with some quantum theory only works to determine intensity for large n. This utilization of classical with quantum theory when they overlap to make progress in quantum theory is known as the *correspondence principle* and was originated by Bohr.

In contrast quantum theory says photon emission and absorption by an atom only occur when electrons transition between orbits. Heisenberg decided first to modify the basic expressions for classical frequency and amplitude to conform to quantum theory. Frequency is rewritten like this:

Classical Frequency

$$f(n,\tau)$$

Quantum Frequency

$$f(n;n-\tau)$$

In classical theory the frequency $f(n,\tau)$ of an emitted light wave of a particular frequency is a function of the orbit n and the τth simple wave which is part of the complex light wave emitted from the electron which is believed to be in a single orbit. In quantum theory the frequency of the light emitted for a particular frequency in the spectrum is a function of the transition of the electron between orbit n and orbit $(n-\tau)$ or $f(n;n-\tau)$. τ in this quantum version of frequency is an integer that must be smaller than n.

Likewise amplitude is modified as shown here:

Classical Amplitude

$$Q(n, \tau)$$

Quantum Amplitude

$$Q(n; n - \tau)$$

In classical theory for large **n** we know that intensity is the absolute value of the amplitude of the simple wave squared as shown here:

Classical Intensity

$$\left| Q(n, \tau) \right|^2$$

In quantum theory the amplitude is believed to be the result of the number of electron transitions between particular orbits (such as in a sample of hydrogen gas that is heated and/or cooled…more atoms experiencing electron transitions equates to greater amplitude and ultimately intensity).   For this reason Heisenberg modified the classical description of a simple wave from this:

$$Q(n, \tau) e^{i 2\pi f(n, \tau) t}$$

To this:

$$Q(n, n - \tau) e^{i 2\pi f(n, n - \tau) t}$$

Heisenberg was trying to move from a wave theory of light to a more particle or quanta theory of light.   Planck and Einstein had led the way in demonstrating that light was made up of particles (as opposed to waves) with energy equal to $h\,f$ for each particle or quanta.   Intensity would then clearly be the total number of electron transitions (in a sample of gas like hydrogen) at a particular frequency multiplied by the energy of a single light quanta $h\,f$.

The problem Heisenberg faced was that q in the Fourier formula represents the position of the electron in classical physics as reflected in the position of the complex light wave at any point in time t.   But Fourier is used to break the complex wave into simple integral frequencies each of whose amplitudes can be determined from the Fourier equation.   These amplitudes when squared give intensity but only when n is large.

So what does q represent in quantum physics?   At this point Heisenberg made a bold move. He took the classical Fourier physics formula describing the position of the electron and modified it to reflect the transition of electrons between orbits rather than describing position on a continuous wave.   The classical Fourier formula for position based on continuous waves is familiar to us as shown here:

$$q = \sum_{\tau} Q(n,\tau)e^{i2\pi f(n,\tau)t}$$

Heisenberg modified this to:

$$q = \sum_{\tau} Q(n,n-\tau)e^{i2\pi f(n,n-\tau)t}$$

The quantum Fourier formula is focused on the transition of the electron from orbit n to orbit $n - \tau.$

This quantum version of Fourier would describe position from a quantum perspective.

Heisenberg knew that at least for n is large the two formulas would give the same answer according to Bohr's correspondence principle.

Next Heisenberg inserted quantum position q into the equation of motion for harmonic oscillation.   The reason he did this was because he knew the forces acting on the electron as it orbited the atom's nucleus were similar to the forces acting on an object suspended from an oscillating spring.   Basic physics says that if you know the forces acting on an object as well as the object's mass then you can determine the exact position of the object at any time.

If q in the Fourier equation represents the theoretical position of the electron based on its classical repetitive wave behavior then putting q into Newton's equation of motion should allow us to understand the exact motion of the electron including when it transitions between orbits.   An object suspended from a oscillating spring gradually oscillates smaller distances as its energy declines and this would be similar to an electron falling into lower energy orbits (and emitting light waves as it transitions from higher to lower energy orbits since electrons have higher energy in orbits further out from the atoms nucleus).   Heisenberg followed this line of reasoning until he eventually derived the equation below:

$$\left|Q(n;n-1)\right|^2 = \frac{h}{8\pi^2 mf} \cdot n$$

With the above equation Heisenberg had finally derived a formula that correctly calculated the intensity of light in atomic spectra that used the key quantum factors of h (Planck's constant) and n (orbit number of electron) that worked whether n was large or small.   This was the major breakthrough in quantum mechanics that allowed physicists to predict the **frequency *and* intensity** of all the atomic spectra.

The above derivation is actually a simpler version of what Heisenberg accomplished.   In his final version he used a more complicated type of oscillation called anharmonic oscillation rather than harmonic oscillation.   This version required an unusual type of math which Heisenberg had to create on his own.   One of the puzzling features of this math was that it was noncommutative which simply means that A times B does not necessarily equal B times A.   Or A plus B does not equal B plus A.   A simple example of this is seen here: **EA+T $\neq$ T+EA**.   That is: **(EAT) $\neq$ (TEA).**

On returning to his home University of Gottingen Heisenberg excitedly showed his results to his colleagues Max Born and Pascual Jordan.   Working together they attempted to come up with an easier way to present Heisenberg's discovery.   During this process Born realized the unusual math that Heisenberg had created was a version of a somewhat obscure type of math known as matrix math.   Amazingly Heisenberg had recreated a version of this branch of math without realizing that it had already been developed!   With this understanding the team soon came up with a completed paper explaining Heisenberg's discovery.   Shortly after it was published in scientific journals.   Heisenberg's solution to quantum mechanics has become known as *Matrix Mechanics*.

For his achievement in solving quantum mechanics Heisenberg received the Nobel Prize in physics for the year of 1932.   Bohr, Born and Jordan also received Nobel Prizes for their contributions to quantum mechanics in subsequent years.

The Photoelectric Effect and Wave/Particle Duality

At about the same time Heisenberg developed matrix mechanics another puzzle involving the photoelectric effect and the wave and particle nature of light and matter was resolved.   Einstein had explained the photoelectric effect as due to light consisting of particles even though most evidence indicated that light consisted of waves.

The photoelectric effect occurs when light beams of varying frequencies are directed at a metal object and this causes electrons to be ejected from the surface of the metal.   It was found that no matter what the intensity of the light that as long as the light was of a particular frequency then every electron ejected had exactly the same energy.   This was a puzzle because it would seem that the greater the intensity of the light at a given frequency then the greater the energy of the ejected electrons.   Think of a water wave with a higher crest (greater intensity) but the same length as a wave with a lower crest.   But Einstein demonstrated that Planck's formula

$$E = hf$$ meant that light behaves as a particle where each quanta of light, i.e. the photon, has

an amount of energy equal to hf .   Planck's constant h means that for a given frequency f that the term hf is a fixed amount of energy for photons of that frequency. This explained why all the electrons ejected by light of a given frequency all had exactly the same energy regardless of the intensity (volume of photons) of the light. Only one photon of fixed energy at a given light frequency would hit an electron at a time and thus the energy of the ejected electron would always be the same at that light frequency.

After learning of Einstein's idea that light can act as either a wave or a particle the French physicist Louis De Broglie proposed that electrons, although considered particles, could also behave as waves! Here is how DeBroglie reached his conclusion outlined step by step:

1) Einstein had equated matter particles to energy in his famous formula: $E = mc^2$

2) Similarly Planck had equated force particles (i.e. Photons) with energy in this formula:
$E = hf$

3) Although photons have no mass they do have momentum which is what allows them to knock electrons off metallic atoms in the photoelectric effect. Momentum $p = mv$. (I.e. mass multiplied by velocity).

4) But the velocity of a photon is equal to the speed of light or "c". If we rewrite Einstein's formula $E = mc^2$ as $E = mcc = (mc)c$ the formula remains the same. But now we can see that the factor "mc" is the same as momentum "p" for light. We can rewrite $E = mc^2$ then as $E = pc$.

5) But Planck's formula also gives energy E. $E = hf$. So we can equate Planck's formula to the modified version of Einstein's formula like this:

$$E = pc = hf \quad \text{or} \quad pc = hf$$

6) Rearranging the formula in step 5 we get:

$$p = hf / c$$

7) This can be rewritten as:

$$p = \frac{h/1}{c/f} = \frac{h}{c/f}$$

But $c / f = \lambda$ (λ = wavelength)

So $p = h/\lambda$

After deriving the final formula in step 7 above DeBroglie proposed that the result demonstrated that momentum "p" which is a characteristic of *particles* is related to wavelength "λ" which is a characteristic of *waves*!

Photons were originally thought to have only wave characteristics until Einstein used the photoelectric effect to demonstrate their particle (momentum) nature.   Now DeBroglie was saying that electrons which were originally thought to have only particle characteristics should also have wave characteristics!

The Uncertainty Principle

Heisenberg made further contributions to quantum mechanics with his now famous ***uncertainty principle***.   The uncertainty principle states that it is impossible to determine various types of measurements of subatomic particles with complete accuracy because of the quantum nature of the universe. For example you cannot determine the precise position and the exact momentum (momentum equals mass multiplied times velocity) of a particle since whatever instrument you use to measure the momentum will change the particles position, and any instrument that is used to determine the position will affect the particles velocity and thus its momentum. The uncertainty principle is important since it explains many mysteries about subatomic physics that would otherwise have remained unresolved.

Heisenberg derived his uncertainty principle by closely examining the puzzling wave and particle duality of nature that Einstein and DeBroglie had demonstrated for light waves/particles and matter waves/particles.   Particles are well defined in that they exist as a point in space taking up a tiny volume which can be measured.   In contrast a wave has no defined location but instead is spread out over a region of space.   However a wave does have definite direction and as we saw with the photon it carries momentum too.

Heisenberg wondered how he could redefine a wave so that it would resemble a particle and in this way resolve the difference between waves and particles.   He realized a mathematical technique called Fourier analysis might just do the trick.

The Fourier analysis uses a combination of waves that interfere with each other to create a well defined wave 'packet'.   A single wave looks like this:

But a collection of different waves combined together via the Fourier analysis looks like this:

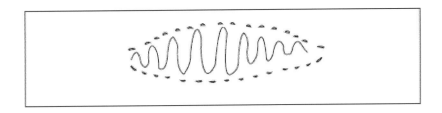

The wave packet created by this technique locates the 'wave' in a well defined position. However Heisenberg realized that a price is paid for obtaining a more accurate position for the wave.   The original single wave had a well defined momentum unique to that wave.   But the wave packet consists of multiple waves each with their own specific momentum.   So in order to narrow down the location of the wave using a wave packet you end up losing precision on the momentum.   Now you have a range of momenta p that we can represent as Δp (pronounced delta p and representing the difference between the beginning and end of the range of momenta).

Likewise with the wave packet.   We know the uncertainty in the position of the wave packet is in a volume given by the range Δx.   Mathematically Heisenberg showed that neither Δx or Δp could be reduced to zero to give the exact position or momentum of a subatomic particle. He demonstrated that the product of Δx times Δp can never be less than Planck's constant divided by 2π or $h/2\pi$.   Physicists show this as $\hbar$ which is pronounced as "h-bar" since it appears often in calculations.   Mathematically Heisenberg's uncertainty principle is written as:

$$\Delta x \cdot \Delta p \geq \hbar$$ (meaning difference in x times difference in p is greater than or equal to h-bar)

The amazing thing about the uncertainty principle is that if you were able to reduce the uncertainty of an electron's momentum Δp to zero then there is no way to determine the location of the electron which could be anywhere in the entire universe!   You can easily see that if Δp equals zero then Δx (position) becomes infinite!

Besides position and momentum other physical variables in nature, such as energy and time, are also subject to the uncertainty principle.   These pairs of variables that are subject to the uncertainty principle are known as 'conjugate variables'.   An interesting consequence of the conjugate variables of energy and time is the production of fermion virtual particles such as electrons.   Although the conservation laws conserving properties such as total electric charge in the universe, total spin, etc. apply to fermions it is possible for virtual fermion particle/anti-particle pairs to come into existence for very short periods.   For example if the time Δt is very short then the energy ΔE can be large enough to create an electron and anti-electron pair pair as virtual particles.   These virtual particles are constantly being produced and then annihilated around us all the time.   What we think of as empty space is actually filled with these virtual particles. These virtual particles play an important role in the interaction of charged particles (such as a real electron and proton) with one another in that they are responsible for what we think of as electric fields, magnetic fields etc.   The virtual particles 'carry' the repulsive or attractive force between charged particles in the form of virtual photons.

Wave Mechanics

The Austrian physicist Erwin Schrodinger took note of DeBroglie's idea that electrons can behave as waves as well as particles.   It therefore seemed to him that the wave nature of electrons might explain the orbital behavior of electrons in an atom.

If the smallest orbit an electron could have around the nucleus consisted of a single wave with its up and down segments then the electron could not spiral into the nucleus since it could not take up an orbit less than one wavelength long. Other orbits further out could also only have a whole number of wavelengths which would explain why the electrons in the various shells could only be at specific distances form the nucleus. It would also explain why the electrons could only have certain spins, tilts, ellipses, etc. Using the wave nature of electrons and the conservation of energy law, Schrödinger worked out a number of formulas to explain electron behavior in 1926.   His mathematical treatment is now referred to as Schrödinger's equations or wave mechanics.   Paul Dirac also contributed to this work.

Below is the derivation of the basic Schrodinger equation.   As promised in the foreword of the book I want to show at least the basic math used in these fundamental physics theories. Readers will get a much better grasp of what quantum theory is all about.   The math is not meant to be rigorous but rather just to give a feel for the concepts involved.   However the math isn't necessary to understand the overall theme of the book.   For general readers you may simply want to scan the Schrodinger equation math and move onto the conclusion of this chapter which is sub headed "Unanswered Questions of Quantum Mechanics".

The derivation is done in a format using a larger font to make it easier to see how Schrodinger derived this equation.   It's not that hard so dive in and check it out!   But readers not familiar with calculus should review the explanation of basic calculus and vector calculus in Appendix A.

Deriving Schrodinger's Equation

(Schrodinger reformulated a basic wave equation in order to derive the three components of the conservation of energy law [total energy=kinetic energy + potential energy] in terms of wave functions.   He wanted to create a new version of the conservation of energy law that could be used to describe the electron as a wave instead of as a particle.)

$$\Psi(x,t) = Ae^{i2\pi(x/\lambda - ft)}$$

Basic wave function for sinusoidal wave moving along **x** axis at time **t**. This gives the amplitude Ψ of the wave at time **t** at position **x** on the **x** axis. The capital letter **A** on the right is the maximum amplitude of the wave.

For simplicity we'll use just Ψ instead of $\Psi(x,t)$ during the derivation.

$$\lambda = h/p$$

wavelength λ equals Planck's constant **h** divided by momentum **p**

$$E=hf \quad \text{or} \quad f = E/h$$

energy **E** of wave is Planck's constant **h** times frequency **f**

Substitute for **f** in wave equation. Substitute also for λ

$$\Psi = Ae^{[i2\pi(xp/h - Et/h)]}$$

$$\Psi = Ae^{(i2\pi xp/h - i2\pi Et/h)}$$

exponent modified to prepare for differentiation

$$\Psi = \frac{Ae^{i2\pi xp/h}}{e^{i2\pi Et/h}}$$

format changed

$$\Psi = Ae^{i2\pi xp/h} \cdot e^{-i2\pi Et/h} \qquad \text{format changed}$$

Now take partial derivative of wave equation with respect to time

$$\frac{\partial \Psi}{\partial t} = Ae^{i2\pi xp/h} \cdot -i2\pi E/h \cdot e^{-i2\pi Et/h} \qquad \text{partial derivative}$$

with respect to time **t**

$$\frac{\partial \Psi}{\partial t} = -i2\pi E/h \cdot Ae^{(i2\pi xp/h - i2\pi Et/h)} \qquad \text{simplified}$$

$$\frac{\partial \Psi}{\partial t} = -i2\pi E/h \cdot \Psi \qquad \text{replace wave function expression}$$

with wave function symbol Ψ

$$\frac{\partial \Psi}{\partial t} \cdot \frac{h}{2\pi} \cdot -\frac{1}{i} = E\Psi \qquad \text{move terms to other side}$$

Multiply **-1/i** times **i/i**

$$-\frac{1}{i} \cdot \frac{i}{i} = \frac{-i}{i^2} = \frac{-i}{-1} = i \qquad \text{having fun with imaginary number } \mathbf{i}$$

Then:

$$E\Psi = i\left(\frac{h}{2\pi}\right) \cdot \frac{\partial \Psi}{\partial t}$$

substitute **i** for **-1/i**

But $\quad \dfrac{h}{2\pi} = \hbar \quad$ physicists shorthand for the common

expression **h/2π** which is pronounced "**h-bar**"

So we get:

(A) $\quad E\Psi = i\hbar \dfrac{\partial \Psi}{\partial t}$

We have completed taking the partial derivative of the basic wave equation with respect to time **t** resulting in formula **(A)** above. What we've done is replaced the variable **E** for energy (multiplied times the wave function) with a constant multiplied by an operator of differentiation. You'll see how we will use this shortly.

Next we'll determine the partial derivative of the basic wave function in respect to position **x** of the wave function along the **x** axis. We'll skip redundant steps that have already been covered in the partial derivative of the wave function with respect to time that we did above.

$$\Psi = Ae^{i2\pi xp/h} \cdot e^{-i2\pi Et/h}$$

basic wave function in a format that we can take the partial derivative with respect to **x.**

$$\frac{\partial \Psi}{\partial x} = Ae^{-i2\pi Et/h} \cdot i2\pi p/h \cdot e^{i2\pi xp/h}$$

partial derivative of wave equation with respect to **x.**

$$\frac{\partial \Psi}{\partial x} = i2\pi p/h \cdot Ae^{(i2\pi xp/h - i2\pi Et/h)}$$

simplified

$$\frac{\partial \Psi}{\partial x} = i2\pi p/h \cdot \Psi$$

replace wave function expression with wave function symbol Ψ

$$\frac{\partial \Psi}{\partial x} \cdot \frac{h}{2\pi} \cdot \frac{1}{i} = p\Psi$$

move terms to other side

Multiply **1/i** times **i/i**

$$\frac{1}{i} \cdot \frac{i}{i} = \frac{i}{i^2} = \frac{i}{-1} = -i$$

having fun with imaginary number **i** again!

Then:

$$p\Psi = -i\left(\frac{h}{2\pi}\right) \cdot \frac{\partial \Psi}{\partial x}$$ substitute **-i** for **1/i**

(B) $$p\Psi = -i\hbar\frac{\partial \Psi}{\partial x}$$

We have completed taking the partial derivative of the basic wave equation with respect to **x** resulting in formula **(B)** above.   What we've done is replaced the variable **p** for momentum (multiplied times the wave function) with a constant multiplied by an operator of differentiation.   You'll see how we will use this too shortly.

Schrodinger now had a new way to describe energy and momentum (each multiplied by the wave function) in terms of a constant multiplied times an operator of differentiation.   He had found that a particle's energy multiplied by the wave function equaled a constant (imaginary number **i** times **h-bar**) times the operator of differentiation of the wave function.   A similar relation held true for momentum.

Schrodinger's goal had been to understand the behavior of the electron in terms of waves rather than the complicated matrix mechanics of Heisenberg.   So now he had two fundamental pieces of the puzzle, energy and momentum, that could be explained in terms of a constant and a term involving waves.

The next step would be to input these new terms into the basic equation covering conservation of energy.   Here is that formula:

$$E = \frac{p^2}{2m} + V$$

E=total energy of the particle

$$\frac{p^2}{2m} = \text{kinetic energy of particle (from motion)}$$

V=potential energy of particle
m = mass of particle

Schrodinger realized he could replace energy **E** and momentum **p** with the new terms he had derived as follows:

(1) $E = E\Psi = i\hbar\dfrac{\partial\Psi}{\partial t}$

The final term on the right becomes the new way to express **energy E** in terms of waves (in an operator of differentiation of the wave function).

$p = p\Psi = -i\hbar\dfrac{\partial\Psi}{\partial x}$

The final term on the right becomes the new way to express **momentum p** in terms of waves (in an operator of differentiation of the wave function).

However since the momentum **p** is squared in the conservation of energy formula, we have to take the square of the term on the right which means the differential operation is applied twice. This looks as follows:

$$p^2 = (p\Psi)^2 = (-i\hbar)^2 \frac{\partial^2 \Psi}{\partial x^2} = -\hbar^2 \frac{\partial^2 \Psi}{\partial x^2}$$

But so far we have been showing momentum **p** in one direction only, the **x** axis. In reality electron waves are three dimensional so we need to take into account the **x, y** and **z** axis.   This means the operator of differentiation on the right for **x** only must be replaced by a three dimensional version like this:

$$\left(\frac{\partial^2}{\partial x^2} + \frac{\partial^2}{\partial y^2} + \frac{\partial^2}{\partial z^2} \right)\Psi = \nabla^2 \Psi$$

Note the short hand term on the right for carrying out the differentiation operation twice.   It's a lot easier than writing out the left hand term!   Also it's important to note that this is now a **vector** expression since it requires more than one coordinate to define position.

Putting this 3-D term into our original momentum **p** squared above we get:

$$p^2 = (p\Psi)^2 = -\hbar^2 \nabla^2 \Psi$$

The full expression for kinetic energy is shown below and the **p** squared result above is inserted into the kinetic energy expression:

$$(2) \quad \frac{p^2}{2m} = -\frac{\hbar^2}{2m} \cdot \nabla^2 \Psi(r,t)$$

Note that I'm showing r in place of x for the variables of the wave function differentiable operator. "**r**" represents position of the electron in terms of the **x, y** and **z** coordinates we are now using.

So with equations **(1)** and **(2)** above we now have new expressions for the particle's total energy and its kinetic energy based on waves. The only term left in the equation for conservation of energy is the term **V** for potential energy. We don't have anything derived from the basic wave equation we started with for potential energy as we did for total energy **E** and momentum **p**. So we'll just modify potential energy **V** in a way similar to what we did for energy and momentum by multiplying it by the basic wave function. The new version of potential energy is this:

$$(3) \quad V = V(r,t)\Psi(r,t)$$

All we've done is multiply the potential energy function times the basic wave function. This makes it similar to the term on the left side of equations **(A)** and **(B)** that we did for energy **E** and momentum **p.**

Well we've done it! We have all the components to rewrite the equation for conservation of energy for a particle in **terms of waves.** Below is Schrodinger's equation with the conservation of energy law terms for total energy, kinetic energy and potential energy replaced by the terms we derived in equations **(1)**, **(2)**, and **(3)** above:

$$i\hbar \frac{\partial}{\partial t}\Psi(r,t) = -\frac{\hbar^2}{2m} \cdot \nabla^2\Psi(r,t) + V(r,t)\Psi(r,t)$$

Schrodinger's Equation

See that wasn't so hard, was it?

There are several versions of the Schrodinger equation but this is the basic non-relativistic one found in most textbooks.   The basic equation is meant for situations involving a single particle in three dimensions.   It can be used for example to describe the behavior of the electron in a hydrogen atom where only one electron is involved.

The Unanswered Questions of Quantum Mechanics

Many questions remain unanswered in quantum mechanics.   For example how can quantum entities, including both matter and force particles, behave as both waves and particles?   The mathematics of both the wave and particle behavior of quantum entities has long since been worked out.   Experiments have completely confirmed this strange wave/particle duality yet the origin of this duality remains a mystery.

Likewise the puzzling quantum behavior referred to as ***non-locality*** cannot be explained by modern physics.   Somehow quantum particles appear to communicate with each other instantaneously regardless of how far apart they are.   The ***exclusion principle,*** formulated by the Austrian physicist Wolfgang Pauli, requires that two electrons in the same orbit must have opposite spin.   If you separate these electrons from each other, even miles apart, they remain ***entangled*** and will continue to spin in opposite directions.   If one electron spontaneously changes its spin then the other will change instantaneously in order to maintain opposite spin.   Somehow these quantum particles communicate instantaneously.   This defies a fundamental postulate of special relativity that nothing can go faster than the speed of light.

Virtual particles are another enigma of quantum theory.   Although they are predicted by the math of the uncertainty principle it raises the question of just how real any "particles" are.   The English physicist Paul Davies has suggested that our understanding of nature at the quantum level can only be based on analogies to what we observe at the macro level since we cannot see or touch quantum particles but can only observe their effects in experiments.

The resolution of these and other questions in quantum mechanics and relativistic physics will, I believe, lead to an understanding of the UFO phenomenon. We will look at an exciting potential explanation of these quantum and other puzzles in the concluding chapters of the book.

The Standard Model
Chapter 7 in Section II

Overview

The Standard Model represents our best understanding of the quantum microworld and has passed every experimental test devised to date.   It is formulated in terms of the interactions among quantum fields such as the electric field, magnetic field, color field, etc.   Because some particles travel at or near the speed of light, special relativity must also be taken into account.   It took roughly from the 1930's to the late 1970's to complete the group of theories that comprise the Standard Model.

Of the four primary forces known the Standard Model explains three of them to a very high degree of accuracy.   These are the strong nuclear force, the weak nuclear force and electromagnetism.   Only gravity is not explained by the Standard Model.

The fundamental constituents of the Standard Model are the matter and force particles already described in chapter 5, The Building Blocks of Our World.   As previously mentioned the matter particles are called fermions and the force particles are known as bosons.   Some of these particles are familiar to us such as the electron in the fermion family and the photon in the boson family. How these particles interact and create the world we see around us is the essence of the Standard Model.

Understanding the historical development of the Standard Model is important since many of the techniques used in its development could be useful in the quest for a final theory of everything, i.e. all the fundamental forces including gravity.   Ultimately the UFO question will be solved from this pursuit of a complete understanding of nature at its deepest levels.

The rest of this chapter explains how the Standard Model developed and the fascinating math and field theories that allowed scientists to unravel some of nature's best kept secrets.   Let's get right into it.

Electromagnetism and Local Gauge Invariance

Physicists realized that when they were smashing subatomic particles into each other in high energy particle accelerators that they were recreating on a small scale the high energy conditions of the early universe.   This meant that they might be able to see some of the forces of nature in their theoretical unified state when the universe was very young.

Prior to the development of particle accelerators there were already hints that the forces of the universe were once united instead of the four separate forces we see today.   The first understanding of force unification occurred with James Clerk Maxwell's theory of electromagnetism in 1864.   Before Maxwell's theory it was thought that the electric force and magnetic force were two separate forces.   But Maxwell proved mathematically that the electric and magnetic forces were two aspects of a single force called electromagnetism.   His equations showed that electric and magnetic fields interacted with each other as they moved through space and time.   The equations predicted that the combined field moved at the speed of light and was in fact the source of light waves as well as other electromagnetic radiation (infrared, X-ray, etc) that is not visible to the unaided human eye.   The photon is the quanta or quantum particle of electromagnetic radiation.

The theory of electromagnetism is known as a ***gauge symmetry theory***. Gauge theories have been crucial to our progress in understanding nature. For this reason an explanation of gauge symmetry theories is essential in order to see how physics developed over the past century.

Gauge symmetry was first introduced by the German physicist Hermann Weyl soon after Einstein published his general theory of relativity in 1915. Weyl was interested in unifying Einstein's theory of gravity and Maxwell's theory of electromagnetism. At the time only these two forces were known to science, the strong and weak nuclear forces were discovered later.

So how does gauge symmetry work? The basic idea is that if you have something like a ruler that you use to measure things then the length (i.e. the gauge) of the ruler should not change as you move the ruler from point to point. In physics the term symmetry is used to describe something that does not change after some operation has been done such as moving an object from one place to another or rotating it in space, etc.

There are different aspects of gauge symmetry. Changes that are the same everywhere in a given region are called ***global gauge changes***. For example if every piece on a checkerboard is moved one space to the right the relationship of the pieces to one another does not change. This is a global change symmetry. However if every piece on the board is moved randomly in different directions this is called a ***local change***. The relationship of the pieces to one another has changed so the symmetry has been lost. But if some unknown force spontaneously occurred to the now unsymmetric checkerboard pieces that restored them to their original spatial relationship to one another this would be called ***local gauge invariance***. The "unknown" force that restores symmetry to the system (the checkerboard) is called a gauge field and has an associated gauge particle.

Local gauge invariance is a fundamental characteristic of the theory of electromagnetism. An example will illustrate this. Let's say we have a number of electric charges, both positive and negative, and these charges are of different voltages and are moving around in a box. If we added say 10 volts to every charge the relationship or potential voltage (electrical) difference between the charges does not change. This would be called a global change. But if we changed each electric charge by a different amount we would find the electric potential differences between the charges has changed. This means the electric charge by itself is not locally gauge invariant. However this is not the end of our story. Something quite remarkable happens. Moving electric charges create magnetic potentials that will exactly compensate for the changed electric charges. So the combined electric and magnetic fields are locally gauge invariant. The quantum particle that carries the information about the changes is of course the photon. A quantum particle to transmit information about changes in combined quantum fields is required whenever there is local gauge invariance. So Maxwell's theory of electromagnetism, the combination of two force fields, is defined as locally gauge invariant.

Hermann Weyl hoped to use this concept of local gauge invariance, where two combined fields mutually compensate for changes in the other field, in his attempt to unify gravity and electromagnetism. Unfortunately the idea did not work for the unification of Einstein's gravity and Maxwell's electromagnetism.

Nevertheless the local gauge invariance concept continued to be useful. For example particle physicists had been studying the interactions of photons and electrons for many years. How atoms and molecules interact with each other and with different types of EM radiation as well as the origin of EM radiation is directly related to the interaction of photons and electrons. Electrons form the outer shells of atoms and the electrons interact with each other via the virtual photons that result from the Heisenberg uncertainty principle. The study of the interaction of photons and electrons is at the heart of everything we understand about chemistry and much of physics.   The body of knowledge associated with the study of interacting photons and electrons is known as *quantum electrodynamics* or **QED** for short.   QED has been found to be locally gauge invariant just like Maxwell's theory of electromagnetism.

You may wonder why there appear to be two theories involving electromagnetism, i.e. Maxwell's and QED.   How come?   Because the quantum world has both a wave and a particle nature.   Wave-particle duality.   Maxwell's equations focus on the wave aspects while QED focuses on the particle aspect.

In addition to local gauge invariance the theory of special relativity is a factor in QED as well. Because photons travel at the speed of light and subatomic particles that photons interact with can also travel at sizeable fractions of light speed it was necessary for physicists to incorporate special relativity into QED theory.

But this too is not the end of the story.   In the early 1950's physicist Chen Yang working at the Brookhaven National Laboratory teamed up with physicist Robert Mills to try and develop a locally gauge invariant theory of the strong force.   We'll see how they expanded on the gauge symmetry concept to include some new mathematical ideas.

The Strong Force, Local Gauge Invariance and Group Theory

In a local gauge invariant theory some physical quantity is preserved after changes in a system. For Maxwell's theory the overall electromagnetic field of a particular system does not change despite changes to the electric component of the combined field.   The changing magnetic potential compensates exactly for changes in the electric potential.   For other fundamental forces in nature besides electromagnetism local gauge invariance can preserve other things such as certain quantum characteristics.

Yang and Mills decided to use isotopic spin (or isospin) as the quantity that would be preserved in the strong force interaction among nucleons.   Isospin was introduced by Werner Heisenberg as a way to distinguish between the proton and neutron which were virtually identical except for their electric charge.   Heisenberg imagined that there was something akin to an arrow pointing up in the proton with a isospin value of **+1/2** and the neutron with an arrow pointing down with a isospin value of **-1/2** as shown in the illustration below.

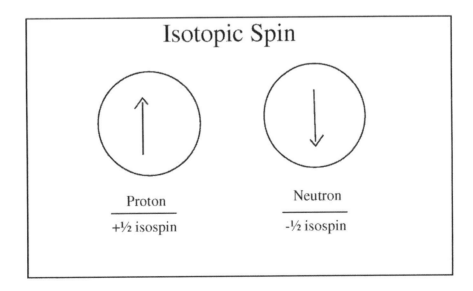

This isospin has no relation to the quantum spin of the fundamental particles and nor does it involve angular momentum. Instead it is a device physicists use to distinguish protons and neutrons as different types of nucleons. Physicists cannot "see" the quantum world so they often have to work by analogies. Physicists imagine the proton rotating in a "internal isospin space" by 180 degrees so it is then transformed into a neutron. Likewise a neutron can rotate by 180 degrees in the internal isospin space to transform into a proton.

Yang and Mills performed these isospin rotations mathematically using a branch of math called **group theory**. Group theory had originally been developed by the Norwegian mathematician Sophus Lie in the 19th century and groups are called Lie groups in his honor. Physicists became interested in this math technique when it was discovered that groups could be used to describe all possible interactions among particular families of subatomic particles. The math restrictions in Lie groups also applied to particle families and determined what interactions were or were not possible between particles in a given family. This obscure branch of mathematics was found to be a valuable tool for modern physics!

The components of a group are arranged in a NxN matrix. In physics the top row and first column of the matrix are used for the matter (or fermion) particles of a particular family of particles. The interior of the matrix is used for all the force particles (also called exchange or boson particles) that interact with the fermions of a particular family. Groups are labeled to describe the type of group and its size. For example SU(3) stands for the "standard unitary group" with a 3x3 matrix. Another group labeled SU(5) with a 5x5 matrix is illustrated below.

SU(5) Group in Physics

	d^R	d^G	d^B	e^+	$\bar{v}$
d^R	G,γ,Z	G	G	X	X
d^G	G	G,γ,Z	G	X	X
d^B	G	G	G,γ,Z	X	X
e^+	X	X	X	γ,Z	W
$\bar{v}$	X	X	X	W	Z

Exchange Particles (bosons)

X=X particle (postulated boson)
G= gluon
γ = photon
W,Z= exchange particles of weak interaction

Matter Particles (fermions)

d^r = red colored quark
d^g = green colored quark
d^b = blue colored quark
e^+ = positron (positive electron)
$\bar{v}$ = anti-neutrino

All possible interactions among the matter particles shown above are covered by this SU(5) group. Neutrinos (in this case a anti-neutrino) convert by a W particle to an electron (positron) and vice versa. The gluons convert different colored quarks into each other. Newly theorized X particles (not yet discovered) are predicted in this SU(5) group to account for electrons transforming into quarks and vice versa. In fact this ability of group theory to predict the existence of new particles, by filling in any blanks in the matrix, is an astounding feature of how math theories impact particle physics.

Its important to note that many of these particle interactions shown in SU(5) as well as other group theories have been observed in particle accelerators either before they were included in group theories or after they were predicted by group theories.

Yang and Mills used the group SU(2) in an attempt to describe the strong force between nucleons. They used isospin as the conserved quantity and built local gauge invariance into their theory. Sure enough, as required by local gauge invariance, gauge particles popped out of the equations. But instead of one, there were three gauge particles! These were ultimately identified as pi-mesons (or pions) and they carried the strong force between nucleons, holding the protons and neutrons all together in the nucleus. The pion had previously been discovered in particle accelerators but now there was the beginning of a theoretical underpinning.

However there were problems. The Yang-Mills theory predicted three massless particles but quantum theory mandates that short range exchange particles have to be very massive. Yet at the same time the theoretically massless pions predicted by Yang-Mills would meet the requirement of symmetry that is an essential element of gauge theory. So how did the pions gain mass in order to match actual observations of massive pions?

Another odd thing about the Yang-Mills gauge particles was that they interacted with one another, unlike the photons in EM theory which do not interact with each other.

Although Yang-Mills had problems it nevertheless opened up a new way to look for force unification. If the masses of the Yang-Mills particles could be explained and the interaction of the pions with each other be better understood then possibly a complete gauge theory of the strong force between nucleons could be constructed. In fact progress would not be long in coming and that success would lead to even more progress as physicists would also discover a whole new aspect of the strong force.

The Electroweak Interaction and the Higgs Mechanism

The Yang-Mills attempt to create a local gauge invariant theory of the strong force ended up with 6 vector fields ("vector" just means they have direction) based on changes in the isospin as the hadrons transformed from one type of hadron to another. Two of these fields were the normal electric and magnetic fields of the EM force with the photon as the gauge particle. The other 4 vector fields consisted of two sets of 2 fields, each of which via local gauge invariance is associated with a gauge particle. These two new particles were massless like the photon but carry opposite electric charge from each other. In order for the theory to make sense masses were input by hand. Eventually these particles were recognized as pions and it was realized that they were composite particles made up of various combinations of quark and anti-quark pairs. Pions belong to the meson family. They carry the strong force that binds protons and neutrons together in the atomic nucleus.

QED, discussed previously, is what is called an Abelian theory. This just means that the order in which particle transformations take place does not affect the outcome. For example an electron will end up in the same quantum state regardless if it first emits a photon and then absorbs it or absorbs a photon and then emits it.   However other gauge theories of the fundamental forces are non-Abelian meaning that the order of particle transformations does affect the outcome.

Yang-Mills had problems such as the fact that the gauge particles had no mass but the short range strong interaction was required by quantum theory to have a significant mass.   Nevertheless the Yang-Mills concept was quickly applied to other forces in particular the weak interaction.

Beta decay is an example of the weak force interaction.   It involves a neutron transforming into a proton and neutrino and then the neutrino transforming into an electron.   This is better known as radioactive decay.   So with the transformation of the neutron into a proton you first have a (local) change in isotopic spin (isospin) which "disturbs" isospin symmetry as we saw in Yang-Mills (thus creating the compensating pion gauge particle).   But in addition we have the neutrino transformed into an electron so by analogy the isospin of these leptons is locally disturbed or changed as well.   Lepton isospin is called "weak isospin" to distinguish it from the isospin of the hadrons like neutrons and protons.

By the late 1950's and early 1960's a number of physicists were working on creating a Yang-Mills type gauge symmetry version of the weak force combined with QED.   QED is included because it explains the interaction of charged particles via virtual photons.   Since weak force processes such as beta decay involve charged particles like electrons and protons it made sense to include QED in any attempt to create a gauge symmetry theory of the weak force.   So physicists began looking for a way to combine QED and the weak force Yang-Mills type gauge symmetry theory.   This came to be known as the electro-weak theory.

But before physicists could nail down the details of the electro-weak theory they first had to thoroughly understand the weak force itself.   They were in for a few surprises.

It was known that the weak force affects both quarks and leptons.   Quarks are the components of nucleons such as neutrons and protons.   Leptons include electrons and neutrinos.

Normally you would think the weak interaction only occurs within each of the 3 families (or generations) of quarks; ie. (u,d), (c,s) and (t,b).   [up and down, charm and strange and top and bottom quarks].   But not so.   It occurs between families but in an unusual way.   It sees them as (u, d'), (c, s') and (t, b').   The d', s' and b' quarks (with the prime notation) are quantum mechanical quarks that are momentary mixtures of the normal d, s and b quarks.   (Technically this partial mixing of quarks from different families is known as the "mixing angle").

A second odd feature of the weak interaction is that weak bosons only couple to left handed fermions.   Handedness refers to the fact that particles have spin and when a particle is going at the speed of light along its axis of spin it will appear to be spinning either to the right or its left in its direction of travel.   (To see this visually curl your four fingers in both right and left hands but with the thumb in each hand pointing up or perpendicular to the curled fingers.   The thumb represents the direction of motion of the particle and you can see the particle spin [curled fingers] means the particle is either right handed or left handed).

Now the concept of handedness would only seem to make sense when a particle is traveling at the speed of light.   At that speed you cannot get out in front of the particle and thus see its handedness change direction from your new viewpoint.   So since fermions like electrons are not massless and do not go at the speed of light then how can we identify their handedness?   In fact as an electron travels at a sizeable fraction of the speed of light it does exhibit handedness and despite traveling slower than the speed of light handedness still applies.

Physicists were very surprised to find out that the weak force bosons only coupled to left handed fermions!   It had always been assumed that left handed and right handed particles interacted with all bosons with equal probability.   This is called mirror symmetry or sometimes parity.   In any case this handedness preference had to be built into any theory of the weak interaction.   Another important point is that any unified theory of the forces must now include this handedness or chirality in order to accurately reflect nature as we know it.

Now armed with a basic understanding of the weak force and already having a solid knowledge of QED, physicists could finally pursue a Yang-Mills type gauge symmetry theory of the electro-weak interaction if it existed.

By the early 1960's the efforts of physicists such as Julian Schwinger, Sidney Bludman, Sheldon Glashow, Abdus Salam and John Ward had created a preliminary version of the electroweak theory.   After combining QED with the weak force they had fashioned a local gauge invariant Yang-Mills non-Abelian theory that required the existence of four gauge particles. These consisted of four vector bosons, namely a W^+, W^-, Z^0 and a photon.   Unfortunately the three particles associated with the weak force alone, i.e. the W^+, W^- and Z^0 ; all had masses of zero and this could not be correct since they were short range particles so they must have mass. Nevertheless, like the Yang-Mills attempt at a gauge symmetry theory of the strong force, it was clear that the theorists were on the right track.   Fortunately help was not long in coming.

Between 1964 and 1966 the mathematical physicist Peter Higgs had developed a theory that would save the day.   Higgs proposed the existence of a new scalar field that permeates all of space and would give mass to the three vector bosons of the Yang-Mills strong force theory which as we saw also had unexplained zero mass gauge bosons.   Higgs primary interest had been to find a natural way for the Yang-Mills strong force theory gauge particles to acquire mass.   There are a total of 4 Higgs scalar bosons with mass in the theory.   According to the Higgs theory when applied to the three pions of the Yang-Mills theory all three of these vector bosons would gain mass. (Scalar means a quantity that has only magnitude but no preferred direction.   This means that no matter what the 'vector' orientation of a subatomic particle it will still pick up a Higgs mass.)   The fourth scalar Higgs would not be involved but might be detectable depending on its mass which had not been predicted.   The Yang-Mills model of the strong force could finally explain the masses of the pions.

Needless to say it was not long before the Higgs scalar bosons were incorporated into the electroweak theories.   By the late 60's physicists Steven Weinberg, Sheldon Glashow and Abdul Salam had created electroweak theories that also gave mass to the weak force bosons via the Higgs mechanism.

The idea behind the Higgs boson is that it did not affect subatomic particles at earlier stages in the universe's existence when temperatures were much hotter than today.   But as the universe cooled the Higgs scalar bosons manifested themselves and gave mass to various subatomic particles.   The first Yang-Mills type strong force theory and preliminary electroweak theories had gauge particles with no mass since these theories represented the forces before the Higgs particles appeared.   It was clear that the four electroweak force bosons were symmetric and indistinguishable from each other in the early universe until the Higgs bosons broke the symmetry.

Therefore at this earlier era in the universe the weak force and the electromagnetic force were unified into a single electroweak force.   So this indicated that in reality there were only three fundamental forces; namely the strong force, electroweak force and gravity.   This was tremendously exciting since it was apparent that the other forces might also be unified in some way. But before we look at this possibility there was one more Yang-Mills type gauge symmetry theory that gave us an entirely new version of the strong force.

One additional note on the electroweak interaction.   The group theory describing the weak force by itself is SU(2).   The electromagnetic force is described by the group U(1).   In the combined electroweak unification (when the universe was younger and hotter) the group theory is described as SU(2)xU(1).   The combined SU(2)xU(1) group is a local gauge invariant non-Abelian theory where all four bosons; i.e. W^+, W^-, Z^0 and the photon; are massless and interchangeable with one another.   But as the universe cooled and the Higgs field interacted with these bosons the symmetry was broken into the separate SU(2) and U(1) groups we see today. Since the SU(2) group particles (W^+, W^-, Z^0) all gained mass, the SU(2) is not a gauge theory because only massless particles can be gauge particles.   However the U(1) group explaining the photon is a gauge group since the photon remained massless.

Nevertheless the concept of local gauge invariance combined with the Higgs mechanism had finally explained the electroweak interaction and why we see only the EM and weak force separately today.

Quantum Chromodynamics

By the early 1960's it was apparent from particle accelerator experiments that the protons and neutrons (baryons) of the atomic nucleus as well as mesons (intermediate mass particles) were composed of smaller particles.   In 1963 the physicists Murray Gell-Mann and George Zweig independently came up with an explanation of the constituent particles that make up the baryons and mesons.   Gell-Mann called these particles quarks while Zweig named them aces.   The name quarks has become the accepted term.   At the time only three quarks had been identified; the up, down and strange.

Experiments at the Stanford Linear Accelerator in California showed that a particle called the omega-minus was made up of three strange quarks which all appeared to be in the same quantum state.   Quarks are fermions and this means that they obey the Pauli exclusion principle such that no two quarks, let alone three, in the same system can be in the same quantum state.   Quarks do have spin but there are only two spin states, up and down, so it couldn't be spin that made the quarks different since that would explain only two of the omega-minus quarks.   In any case it was found that all three strange quarks in the omega-minus particle had the same spin.

Theorists realized there must be some quantum property that makes the three strange quarks different from each other in the omega-minus.   It was decided to label the unknown quantum property of quarks as 'color charge'.   Each of the strange quarks must, it was reasoned, carry a different color charge.   The colors red, blue and green were chosen as the three colors. Theoretical math considerations had pointed to the idea that quarks combining together into particles had to be 'colorless' as far as the color charge was concerned.   So the primary colors of red, blue and green were selected since these three colors when combined in the art world are considered to produce a colorless white.   A proton consisting of two up quarks and one down quark would have to have all three colors such as a red up, a green up and a blue down quark. When a particle consists of just two quarks then theorists say it consists of a quark and a anti-quark pair so for example a red quark with anti-red quark would be colorless.   It was suspected that this color charge was the origin of the strong force that held the quarks together in baryons and mesons with each quark having a different color.   By analogy electrons and protons have different electric charge which holds them together in the atom.

Just as moving electric and magnetic fields when combined produce a locally gauge invariant field we call the electromagnetic field so by analogy the color fields when combined should produce a locally gauge invariant field or fields as well.   The group theory SU(3) was applied to the color fields since there are three different color charges (or fields).

As expected the Yang-Mills non-Abelian type theory developed by Murray Gell-Mann and Harald Fritzsch generated a total of eight gauge particles representing the color force that binds the quarks together into protons, neutrons and other hadrons (hadrons are particles that feel the strong force…ie. baryons with three quarks each as in protons and neutrons and mesons composed of a quark and anti-quark pair like the pion). These eight exchange particles are called **gluons**. The gluons not only interact between the quarks but also interact among themselves which makes studying the color force interaction a real challenge. By comparison the EM force is represented by the single photon; the weak force is represented by the three exchange particles W^+, W^- and Z^0; and the strong force is represented by the eight gluons.

It quickly became clear that the color charge is the source of the fundamental strong force holding together the quarks making up protons, neutrons and other hadrons. The pi-mesons (pions) of the Yang-Mills theory now could be explained. Pions consisting of quark/anti-quark pairs carry color charge so it is "left over" color charge in the pions that holds protons and neutrons together in the nucleus. In turn, gluons carrying color charge as well hold the quarks together that make up the protons, neutrons and mesons. Protons and neutrons are composed of three quarks each while mesons consist of two quarks each.

The entire theory of how quarks interact with one another via the gluons of the color force is known as **quantum chromodynamics** or **QCD**. The name was chosen to deliberately resemble the quantum electrodynamics (QED) theory that describes how electrically charged particles interact with each other via photons of the electromagnetic force.

Renormalization and the Gauge Force Theories

One other feature of the gauge theories needs to be mentioned. The mathematics of the gauge theories initially produced unacceptable infinities that caused physicists great concern. In quantum electrodynamics (QED) the electron is the source of the electric field. This field drops off in strength by the inverse square of the distance **r** from the electron or $1/_{r^2}$. But since the electric field originates with the electron then theoretically it should interact with itself. Of course the electron is zero distance from itself so by applying the formula $1/_{r^2}$ we get $1/_0$ or one divided by zero. This equals infinity!

Since the strength of the electric field is its energy but energy and mass are equivalent then the electron should have infinite mass! Clearly the electron does not have infinite mass so this was a problem for the early work on QED!

Another example of infinities arising in gauge theories is calculations involving the interaction of quantum particles such as two electrons colliding. Richard Feynman developed special diagrams, called Feynman diagrams, that show how quantum particles interact by detailing all possible interactions. Unfortunately for any particular process there are an infinite number of such diagrams!

With the colliding electrons there can be several photons exchanged that during the exchange convert into electron and anti-electron pairs before reverting back to photons and being absorbed by the electrons. Countless other processes like this occur from the interactions of the two colliding electrons. For each possibility a separate Feynman diagram can be drawn. A process called perturbation is used to calculate the energies involved in these interactions and how they contribute to the masses of the particles involved. It turns out only a few levels of interactions need to be calculated based on the nature of the successive interaction energy contributions. Unfortunately as with the electron's self interactions the perturbation calculations also lead to infinities.

Fortunately mathematical methods were developed to solve the infinities that were popping up in relativistic quantum gauge field theories. A technique called renormalization successfully rid the gauge theories of their infinities. The Feynman diagrams result in infinities for the mass, charge and coupling constants of interacting particles like electrons. But from experiment we know the actual mass, charge and coupling constant of electrons so by incorporating this information in the Feynman process the infinities are vanquished.

Renormalization works with only a selected number of theories but interestingly all the gauge theories are renormalizeable. Is nature trying to tell us something?

The Standard Model: Its Triumphs and Shortcomings

In the early decades of the twentieth century physicists had constructed the two foundations of modern physics with the theories of relativity and quantum mechanics. The gauge field concept was the crucial factor that allowed physicists to make the next major leap in understanding the natural world at its deepest levels. Using the powerful tool of gauge symmetry physicists were able to explain all of the material world in terms of its basic matter particles of leptons and quarks which interact via the gauge boson force particles.

The crowning achievements of the gauge field concept are the electroweak theory and the theory of quantum chromodynamics which were fully formulated by the late 1970's. The electroweak theory explains the electromagnetic force and the weak force and how they interact with other particles to an astonishing degree of accuracy as determined from particle accelerator experiments. Likewise QCD explains the strong force and how quarks and colored gluons interact. QCD has also proved immensely accurate in its predictions. Both theories incorporate quantum mechanics, special relativity and gauge fields in what is referred to as a relativistic quantum field theory. Together these two theories are called the Standard Model and represent one of the greatest accomplishments in scientific history.

But despite its successes the Standard Model has serious shortcomings. Some of the major deficiencies of the Standard Model are listed here.

1) Masses of fundamental particles have to be put in by hand. They don't arise naturally from the theory. There is no particular pattern to the masses of the particles, they are all over the map.

2) Theoretical considerations require the mass of the Higgs boson to be around 114 to 250 GeV. Higgs is responsible for giving the fundamental masses to the various particles. But that raises the question of why the masses of the basic particles are so different.

3) A number of constants such as the coupling constant and mixing angles have to be put into the Standard Model by hand. They do not emerge from the Standard Model by some logical process.

4) Why are there three families of particles among the quarks and leptons? For example the electron, mu electron and tau electron in the lepton family are the three generations of electrons.

5) The Standard Model unifies the electromagnetic force and the weak force plus it explains the strong force. However it does not unite the electroweak and strong force.

6) The mysterious and still undiscovered Higgs particles.

7) Finally the Standard Model does not include gravity so it is clearly incomplete.

Even with its shortcomings the Standard Model is still a tremendous achievement and represents many decades of hard work by many people.   Moving beyond the Standard Model is the challenge facing physicists now and it is here where we will be looking to see how we might solve the UFO enigma.

Grand Unified Theories
Chapter 8 in Section II

Soon after its completion physicists began to look at how they might extend the Standard Model into new areas.   With the Standard Model they had a quantum theory of the strong force, weak force and electromagnetism.   In addition they had a unified theory of two of these three forces with a theory combining electromagnetism with the weak force resulting in the electroweak force theory.   The next logical step would be to combine the electroweak force with the strong force into a grand unified theory or GUT of the three non-gravity forces.   Is there a theory that could incorporate all the electroweak force particles plus the gluon force particles such that it contained all particle physics except for the gravitons?   To put it differently it should include the entire symmetry group of SU(3)xSU(2)xU(1).   SU(3) includes the 3 color charges.   SU(2) includes the W and Z particles.   U(1) explains the photon.

Physicists just had to look at the math of group theory to see that for the GUT to work it required 24 gauge bosons; 12 we know (8 gluons, 2 W's, Z and the photon) and 12 new ones known as the X bosons.   These 12 new X boson particles would have various combinations of color charges and fractional electric charge.

The minimal symmetry group that encompasses these 24 gauge bosons is SU(5).   SU(5) group consists of a header row and first column of five matter particles each.   These matter particles transform into each other via various combinations.   An example would be these 5 matter particles: $(d^r, d^g, d^b, e^+, \bar{v})$.   I.e. three d quarks, an anti-electron plus an electron anti-neutrino. Note that the -1/3 for each of the three d quarks plus the +1 of the anti-electron adds up to zero electric charge (the electron anti-neutrino of course has no charge).   The Standard Model without GUT cannot explain why the d quark is -1/3 electric charge but the GUT explains it!

The SU(5) group theory matrix is shown below.   (Used earlier in Chapter 7 too).

SU(5) Group in Physics

	d^R	d^G	d^B	e^+	$\bar{v}$
d^R	G,γ,Z	G	G	X	X
d^G	G	G,γ,Z	G	X	X
d^B	G	G	G,γ,Z	X	X
e^+	X	X	X	γ,Z	W
$\bar{v}$	X	X	X	W	Z

Exchange Particles (bosons)

X=X particle (postulated boson)
G= gluon
γ = photon
W,Z= exchange particles of weak interaction

Matter Particles (fermions)

d^r = red colored quark
d^g = green colored quark
d^b = blue colored quark
e^+ = positron (positive electron)
$\bar{v}$ = anti-neutrino

In SU(5) there are new interactions with quarks turning into leptons. In the Standard Model there is quark conservation but in the SU(5) GUT this is no longer the case. This means that since a proton is made of quarks there are some SU(5) allowed interactions that would cause a proton to decay into a mixture of other quarks and leptons. I.e. this implies that protons are not stable!

The strength of a force interaction depends on the energy level in the universe at the time the force strength is measured. Also we know that the greater the mass of a force particle the shorter the range of that force. The X bosons are very massive. So if two quarks happen to get within 10^{-31} meters then by the Heisenberg uncertainty principle they can exchange an X boson and cause proton decay! Of course this is a rare event or otherwise we humans would disintegrate in a hurry!

In the early universe when all matter was crammed very close together these interactions of X bosons occurred all the time. But as the universe expanded and cooled these interactions became rare.

Grand unification takes place at an energy of about 10^{15} GeV in SU(5) which is not too far from the Planck energy of 10^{19} GeV. Since X bosons are exchanged at a energy of 10^{15} GeV it means the mass of the X boson is around 10^{15} Gev. This compares to the mass of the W boson which is less than 10^2 GeV.

Unfortunately SU(5) is ruled out by experiments but it did help explain a lot of physics. However several other symmetry groups have been developed as possible GUT theories. These are the groups SO(10) and E(6). Since both these theories predict a much lower probability of proton decay that means they cannot be ruled out by proton decay experiments. These newer GUT theories also help explain additional physics such as explaining why there are three color charges. But despite these theoretical successes there are some serious problems with all GUT's. These are outlined below.

GUT Problems

1) There seems to be no new quantum particles between the electroweak scale of 10^2 GeV and the GUT scale of 10^{15} GeV except for the Higgs at possibly 250 GeV. This compares to 100 GeV (or 10^2 GeV) for the electroweak scale.

2) GUT's explain a number of the parameters of the Standard Model but still many remain unexplained. GUT's cannot explain masses of leptons and quarks nor why there are 3 generations of matter.

3) A bigger problem is when we simulate mathematically an increase in the energy of the universe (to replicate an earlier epoch after the Big Bang), the coupling constants which measure the force interaction strengths for U(1), SU(2) and SU(3) get close at 10^{15} GeV but are not exactly the same. So some other physics must be needed.

4) But the biggest problem facing the GUT's is the gauge hierarchy problem. A fermion's mass is proportional to its coupling strength with the Higgs boson. But since the Higgs is a spin-0 it interacts, according to quantum theory, such that it takes on the maximum mass of whatever particles it interacts with. This is because the Higgs boson as a scalar spin-0 particle not only self-

interacts but interacts with other particles to get its mass.   Since the maximum mass in the GUT is the 10^{15} GeV of the X bosons that means the (electroweak) Higgs should have a mass of 10^{15} GeV. But other measurements of the electroweak scale indicate the Higgs should have a mass in the range of 114 GeV to 249 GeV.   So the problem is the huge discrepancy between the theoretical masses of Higgs – one estimate around 10^2 GeV (ie. the 114 GeV – 249 GeV) and the other Higgs '2' of 10^{15} GeV at the GUT scale.   Again some new physics beyond the Standard Model must come into play.

Supersymmetry and 4-D Supergravity
Chapter 9 in Section II

Supersymmetry

All known laws of physics are not altered by various spacetime transformations such as translation (moving an object from one point to another), rotations (rotating an object in space), and boosts (which changes an objects velocity, ie. accelerates it to a different speed).   The complete set of these spacetime symmetries is called Poincare symmetry after the French mathematician Henri Poincare.   These symmetries allow similar objects to transform into each other such as a neutrino into an electron and vice versa.   The key is that these objects need to be in the same family.   In our example of the electron and neutrino, both particles were in the fermion (matter) family.

Then physicists accidentally stumbled upon a new type of symmetry that goes beyond Poincare.   The new type of symmetry is called supersymmetry and allows fermions to transform into bosons (force particles) and vice versa.   Supersymmetry is the maximum possible symmetry beyond Poincare symmetry that is mathematically consistent.   If it exists it indicates that nature uses all possible symmetries.

Supersymmetry transformations always involve a change of ½ spin which then transforms a boson into a fermion or a fermion into a boson.   For example a matter particle (fermion) such as an electron with spin ½ would become a boson with spin 0 called a selectron.   For Standard Model fermion (matter) particles the supersymmetry bosonic partners add the letter "s" in front.   For Standard Model bosons (force particles) the letter "o" is appended for their fermion 'superpartners'. The photino would be the fermionic supersymmetry partner of the Standard Model bosonic photon as an example.   See the chart on the next page of Standard Model particles and their supersymmetry partners.

Supersymmetry was first developed in the early 1970's by a number of different physicists. In the U.S. Andre Neveu, Pierre Ramond and John Schwarz developed an early version of supersymmetry.   In 1974 Julius Wess in Germany and Bruno Zumino at CERN tied supersymmetry to quantum theories.

Supersymmetry went a long ways toward solving the major problems the GUT theories had such as bringing the coupling constants closer together for the three non-gravity forces at high energy as shown below.   It also theoretically resolved the hierarchy problem.

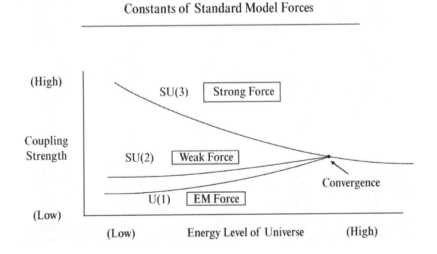

Supersymmetry Effect on Coupling
Constants of Standard Model Forces

Chart of Standard Model Particles and Their Superpartners

	Standard Model Particles		**Superpartners**	
	Name	Spin	Name	Spin

Quarks (matter particles)

	Name	Spin	Name	Spin
	up	$1/2$	sup	0
	charm	$1/2$	scharm	0
	top	$1/2$	stop	0
	etc.	$1/2$	"squarks"	0

Leptons (matter particles)

	Name	Spin	Name	Spin
	electron	$1/2$	selectron	0
	muon	$1/2$	smuon	0
	tau	$1/2$	stau	0
	neutrino	$1/2$	sneutrino	0
	etc.	$1/2$	"sleptons"	0

Bosons (force particles)

	Name	Spin	Name	Spin
	photon	1	photino	$1/2$
	gluon	1	gluino	$1/2$
	Z^0	1	zino	$1/2$
	$W^\pm$	1	wino	$1/2$
	Higgs	0	Higgsino	$1/2$
	graviton	2	gravitino	$1/2$

Another problem with the GUT's was that the Higgs particle, by quantum mechanical effects, would gain a mass of the heaviest X bosons that are needed at the GUT unification scale of 10^{15} GeV. But this contradicted the small mass of about 10^2 GeV for the Higgs that was needed at the electroweak scale. This gave two wildly different values for Higgs and is called the hierarchy problem. However with supersymmetry the new superpartner particles that interact quantum mechanically with the Higgs cancel out the masses from the non-supersymmetry particle interactions with the Higgs so the hierarchy problem disappears.

An exciting feature of supersymmetry is that it is a quantum theory that appears to include gravity. Repeated supersymmetry transformations such as transforming a spin ½ fermion into a spin 0 boson and then back into a spin ½ fermion causes the spin ½ fermion to move to a different spatial position. This is a Poincare transformation which is associated with general relativity, Einstein's theory of gravity!

4-D Supergravity

Supersymmetry could be either local or global. Global symmetry means if we do a supersymmetry transformation at one location we must make the same transformation in all other locations in space. If we do so the laws of physics stay the same.

But what about a local supersymmetry transformation? What if we do a local supersymmetry transformation such as transforming an electron into a selectron? What happens to the laws of physics? The laws of physics will only stay the same if there are compensating force fields created by nature. For example a local non-supersymmetry transformation of the electric field (like an electron transformed into a neutrino) does not change the laws of electromagnetism because a compensating gauge field, represented by the photon, is created by nature. Likewise a local supersymmetry transformation creates compensating gauge fields manifested by the graviton and its superpartner called the gravitino. The gravitino results from a single supersymmetry transformation such as a fermion into a boson. Quantum theory requires this gravitino to have spin of 3/2. The graviton results, as mentioned previously, from two consecutive supersymmetry transformations such as a fermion into a boson and back to a fermion. Quantum theory requires the graviton to be a massless spin 2 particle.

This local supersymmetry theory is called *supergravity*. The earliest supergravity theory was developed in 1976 by Daniel Freedman and Peter Van Nieuwenhuizen. This was a basic supergravity theory that contained just one graviton, one gravitino and no other particles. Of course a supersymmetry theory that contained no fermions or non-gravity force particles could not describe our world but at least it was a quantum theory that included gravity. Developing a quantum theory of gravity is a kind of holy grail for physicists. Einstein's general theory of gravity explains this force in terms of the geometry of spacetime.

In time seven more supergravity theories were developed with ever increasing numbers of particles that included both fermions and bosons. The total of eight different supergravity theories are labeled N=1, N=2, N=3, etc. up to N=8. For example N=3 includes three gravitinos, one graviton, three spin 1 non-gravity force particles and one spin ½ matter particle. The supergravity theory with the maximum symmetry is N=8 with eight gravitinos, one graviton, 28 spin 1 particles, 56 spin ½ particles and 70 spin 0 particles. Theories with N greater than 8 are not allowed because such theories would result in a graviton with spin greater than 2 which does not match our understanding of the graviton.

Supergravity N=8 looked very promising as a quantum theory of everything, including gravity. The theory helps to tame the infinities associated with gravity.   Infinities arise as a problem with any theory of gravity since gravitons interact with each other.   This means whenever you try to describe the gravitational interaction between two objects with mass you have to deal with these graviton infinities.   Fortunately with supergravity N=8 the infinities associated with gravitons are mostly canceled out by the gravitons superpartner gravitinos.

Unfortunately despite the aesthetic beauty of the theory, supergravity has some fatal flaws. The supergravity gauge group O(8) does not include the full Standard Model of SU(3)xSU(2)x U(1). It also does not include all the leptons of the Standard Model.   In addition chirality is missing and we know this is a feature of our world.   Finally the gravitational infinities are not vanquished completely.

Clearly some new physics was needed to explain nature.   Fortunately even while some physicists worked on 4-D supergravity others were looking at a different approach to explaining nature.

5-D Kaluza-Klein Theory and 11-D Supergravity
Chapter 10 in Section II

5-D Kaluza-Klein Theory

In 1919 the German physicist Theodor Kaluza wrote down Einstein's equations of general relativity in 5 dimensions rather than the 4 Einstein used.   To his surprise both Maxwell's 4-D equations of electromagnetism plus Einstein's general relativity can be derived from the 5-D version of general relativity.   This was one of the first hints that the universe may consist of more than 4 dimensions.

The Swedish physicist Oscar Klein noticed that Kaluza's 5-D version of general relativity was a classical theory.   Klein realized he would have to modify Kaluza's theory to take into account quantum mechanics since by the 1920's it was clear that nature obeyed the rules of quantum theory as well as general relativity.   Klein modified Kaluza's idea by reformulating the basic equations of quantum mechanics in 5 dimensions..   Solving these 5-D quantum equations indicated that the results could be viewed as electromagnetic and gravitational waves in 4-D space.   As we saw in the chapter on quantum mechanics the Schrodinger equation describes particles as waves.

Klein realized that from a quantum mechanics perspective the 5th dimension must be small or otherwise we would have noticed it.   For example the 5th dimension has to be smaller than a molecule or we would have noticed some molecules disappearing into the 5th dimension! Likewise for subatomic particles we know from quantum theory that the smaller the wavelength of a particle the greater the energy needed to probe at that short distance.   Since current particle accelerators can probe at distances of about 10^{-18} meters then we know the 5th dimension is smaller than that since no particles have been unaccounted for so far in accelerator experiments.

Although Klein improved on Kaluza's original idea the theory combining electromagnetism and gravity via the use of a 5th dimension is usually referred to as Kaluza-Klein (KK) theory.   KK simplifies our understanding of of these two forces by postulating that there is really only one force, the gravity force.   The electromagnetic force then results from vibrations of the gravity field in the 5th dimension.   Since we are only aware of 4 dimensions it appears as if the gravity and electromagnetism are two separate forces.   Five dimensional spacetime should also create a scalar spin-0 particle called the dilaton, analogous to the way the photon is required in 4-D electromagnetic theory.

Klein postulated that the 5th dimension consisted of a tiny closed loop or circle at each point in spacetime.   How can the circumference of the 5th dimension be determined?   Since the 5th dimension unites gravity and electromagnetism then we know that any particle vibrations manifested around the 5th dimension will have both mass (associated with gravity) and electric charge (associated with EM force).   The simplest vibration would be one complete oscillation of a wave around the 5th dimension.   The next particle would be two complete oscillations.   And so on giving birth to a cascade or tower of particles that would be characteristic of the 5th dimension.   If we assume the electric charge associated with the simplest KK particle with mass is equal to the electric charge of an electron we can calculate the circumference of the 5th dimension.   The circumference is proportional to the inverse of the electric charge.   Therefore the circumference ends up being about 10^{-32} meters or almost the Planck length.   This translates into a mass for the lowest mass KK particle of 10^{16} times the mass of the proton, far beyond our accelerator testing capability.

Despite the impressive accomplishments of KK theory scientists began to have doubts. The dilaton field/particle was estimated to have low mass so it should have been detected in our particle accelerators. Unfortunately there has been no sign of the dilation field or particle. There were other problems too with KK related to the size of the extra dimension that finally caused interest in the theory to wane.

11-D Supergravity

In the chapter on the Standard Model we saw that a local symmetry transformation of the electric field such as converting an electron into a neutrino does not affect the laws of electromagnetism as long as a compensating gauge field/particle is created. This compensating particle is the photon. KK theory says that this invariance is the result of the circular 5th dimension. General relativity is invariant under rotations around a circle too. The mathematical group that describes this general relativity invariance is U(1), exactly the same U(1) invariance that describes electromagnetic invariance.

If it is possible to explain gravity and electromagnetism in 4-D as manifestations of just one force (gravity) in a 5-D universe might it be possible that all the forces could be united by using even more extra dimensions? This turns out to be the case.

In 1978 Eugene Cremmer, Bernard Julia and Joel Scherk in Paris constructed an eleven dimensional supergravity theory with N=1. Amazingly this theory reduced automatically to a 4-D space-time that we experience as well as a curled up 7-D space. In the process of reduction the 4-D space-time then included N=8 supergravity which, as noted before, is the most promising of the supergravity theories. So the Cremmer, Julia and Scherk theory combines N=8 supergravity with the concept of additional dimensions beyond those proposed by KK theory. Could the 11-D spacetime unite all four fundamental forces of nature instead of just the gravity and electromagnetic force as in KK theory? Further developments did indeed point in that direction.

Ed Witten, a mathematical physicist at Princeton, calculated in 1981 that the minimum number of extra dimensions needed to encompass all the symmetries of the Standard Model is seven. The electromagnetic force U(1) needs one extra dimension beyond the 4 dimensions of classical spacetime as KK theory had determined. The weak force SU(2) needs an additional two dimensions. And SU(3), the strong force symmetry, needs four extra dimensions. These add up to a total of seven extra dimensions in order to unite all the Standard Model forces with gravity using the concept of extra dimensions.

So if you can imagine an extra seven dimensions beyond the four of space-time that we are familiar with then the Standard Model symmetry group SU(3)xSU(2)xU(1) would be unified with gravity for a Theory of Everything (TOE). It turns out that a workable supergravity theory can only have a maximum of eleven dimensions or else particles with spin greater than two must exist. But theories of quantum gravity will not work if such particles exist. However the minimum of seven extra dimensions determined by Witten plus the original four dimensions of known space-time equals just eleven dimensions. This is very encouraging since two different methods give the same number of extra dimensions so it appears that physicists are on the right track for a Theory of Everything.

By adding seven curled up dimensions to the known 4-D space-time we get an 11-D supergravity theory that includes all four fundamental forces of nature. Remember that the 4-D (not 11-D), N=8 supergravity theory did not include the full Standard Model symmetry of SU(3)xSU(2)xU(1). It also did not include some of the leptons of the Standard Model. But now it looked like these problems had been solved. Without question physicists had made significant progress toward a unified theory of all the forces in nature.

Unfortunately even 11-D supergravity had flaws. As with the 4-D supergravity, 11-D supergravity does not include chirality. Chirality refers to the fact that weak force bosons only interact with left handed fermions. In addition the infinities associated with the gravity force though less troublesome, still remained a problem. It was clear that further new physics would be needed to resolve the remaining problems. Nevertheless it was also clear that some progress had been made.

Strings & Superstrings and 10-D & 11-D M-Theory
Chapter 11 in Section II

10-D & 26-D Strings

11-D supergravity did not work completely.   But it does appear to work at a particular energy level indicating that this theory represents some earlier stage of the evolution of the universe.

The problem with 11-D supergravity is that it assumes all the fundamental subatomic particles are zero dimensional points.   For interactions involving the graviton this leads to infinities. Other theories such as QED and QCD could renormalize the infinities away but this approach did not work for gravity.   It is at this point that the concept of fundamental quantum particles being made up of tiny, vibrating one dimensional strings entered the picture.

An Italian physicist named Gabriele Veneziano wrote down the equations that described the behavior of hadrons in experiments at CERN during the late 1960's.   Veneziano had noticed that math functions developed by a Swiss mathematician named Leonard Euler some two centuries previously could be used to describe the hadron interactions.   Several years later a number of physicists realized that Veneziano's equations described something akin to tiny wiggling strings. These string theory pioneers were Yoichiro Nambu who resided at Chicago University, Leonard Susskind in New York and Holger Nielson in Copenhagen.   All of these physicists studied the behavior of these hypothetical strings in relativistic interactions.

There seemed to be two types of strings, open or closed.   An open string with two disconnected ends and a closed string with the ends connected forming a string loop.   Electric charge and color charge are located at the end points of an open string.   With closed strings charges are smeared all around the string loop.   The vibrations of these strings corresponded to different fundamental particles.   Calculations indicated the strings would be under enormous tension so they contained tremendous energy.   Theories that have at least some open strings are called Type I while closed string theories are designated Type II by physicists.

Because of the extended size of the strings it was quickly realized that this could eliminate the infinities that had plagued previous theories.   Earlier theories treated the fundamental particles as dimensionless points and this is what leads to infinities.   The strength of the electric field between two charged particles, for example, is given by $1/_{r^2}$.   But if the particles are dimensionless points then **r** can get down in size all the way to zero.   Dividing one by zero squared yields infinities.

Perturbative theory is still used in string theory just as it was necessary in the Standard Model to calculate the probability of a particular string interaction occurring.   Perturbation calculations are easier in string theory for at least the weak interactions but remain difficult for other interactions.

The original version of string theory only described bosonic interactions.   This was remedied in 1971 by physicists Andre Neveu, John Schwartz and Pierre Ramond who developed a version of string theory that incorporated fermions called spinning string theory.   It was also found that string theory only made sense if space-time had more than four dimensions.   Symmetry considerations required that bosonic string theory exist in exactly 26 dimensions.   Spinning string theory required fermions to exist in exactly 10 dimensions.   This was remarkable since it was the first theory of nature that actually specifies the number of space-time dimensions!   However it did seem odd that so many dimensions were needed to explain hadron interactions.

Another problem with string theory included the predicted existence of tachyons, particles which can travel faster than the speed of light.   In addition energy cancellations from quantum mechanical considerations gave the theory lots of massless particles but not the massive particles needed to explain hadrons.   By the early 1970's QCD (quantum chromodynamics) was able to explain the strong interactions far better than string theory and interest in string theory began to wane.

However in 1974 interest in string theory was revived when John Schwarz and Joel Scherk demonstrated that one of the massless string particles was the spin-2 graviton!   This was tremendously exciting because it meant that string theory incorporated general relativity, the theory of gravity.   Schwarz and Scherk then suggested that string theory was being used incorrectly in describing just the strong force which caused physicists to focus only on hadron sized particles of around 10^{-15} meters.   The real focus should have been strings the size of 10^{-35} meters or around the Planck length.   Strings of this size would have the possibility of explaining all the fundamental forces in a true Theory of Everything.

In order for strings to explain particles with low masses, like the electron, it would have to have slightly less than perfect quantum mechanical cancellations.   Massless particles are explained, such as the graviton, by perfect cancellations.   But to this day string theory has not been able to explain the masses of the non-zero mass fundamental particles.   Yet there is hope that all the masses of the quantum particles in the Standard Model can eventually be explained in string theory.

Superstrings

Further advances in string theory came quickly.   In 1976 Scherk, Ferdinando Gliozzi and David Olive showed that spinning string theory could be made supersymmetric with equal numbers of fermions and superpartner bosons.   In addition they were able to eliminate the tachyons.

In the early 1980's John Schwarz along with Michael Green, an English physicist, and the Swedish physicist Lars Brink developed a formal theory of strings with supersymmetry that they called superstring theory.   As mentioned previously there were versions of supersymmetry, the pre-string theory, called supergravity that included the graviton.   Supergravity had multiple versions known as N=1, N=2, etc up to N=8, with N corresponding to the number of superpartner gravitinos.   For superstrings N=1 was found to be a Type I theory while N=2 was a Type II theory.   As with the original string theory, the superstring theory required 10 dimensions, 9 of space and one of time,

But further problems in string theory had to be dealt with in order to make it viable.   Since we only see three large dimensions and one of time, superstring theory had to explain why we don't see the remaining 6 dimensions of the 10 required for superstrings.   This problem is resolved by assuming the remaining 6 dimensions are curled up small dimensions similar to those envisioned in KK theory.

Unfortunately requiring that the remaining dimensions to be small makes it difficult for superstring theory to show that chirality (ie. the handedness of the weak interaction) still occurs in the large 4-D world we see.   But in 1984 Green and Schwarz were able to resolve this problem but only by putting severe restrictions on the allowed gauge symmetry groups to be contained by superstring theory.   The early GUT gauge symmetry of SU(5) includes the combined symmetries of U(1)xSU(2)xSU(3).   Other versions of these gauge symmetry groups were also developed such as SO(10) since SU(5) still had some problems.   SU(5) included 24 gauge bosons while SO(10) included 45 gauge bosons.   Green and Schwarz found that in order to eliminate problems with chirality the superstring theory required either a gauge symmetry group of SO(32) or E(8)xE(8).

At this point superstring theorists had made significant progress in developing a potential unified theory.   The accomplishments can be summarized as follows:

1) Superstring theory required supersymmetry which physicists believe is a part of our world.

2) The infinities that plagued all other theories appear to be tamed.

3) Superstring has just one basic energy scale caused by the string tension.   This causes the theory to be very restricted with virtually no room to adjust constants.

4) The theory predicts the existence of the graviton.

5) Unlike the point particle based Standard Model, superstring theory does not need a huge number of unexplained fundamental particles.   There is just the one fundamental vibrating string whose different modes of vibration account for all the particles we see in nature.

6) Superstring theory predicts a 10 dimensional universe rather than physicists having to input this constant by hand.

7) The gauge symmetry required by our 4-D world is predicted by superstring theory.   Although there are two candidates, SO(32) and E(8)xE(8), this is still very encouraging.

8) Lastly the superstring theory as a unified theory is effective at high energies and small distances (ie. the Planck scale).   But at lower energies we would not be aware that the fundamental constituents of matter are vibrating strings.   This means our other theories such as quantum field theory are still effective theories at lower energies.   Supergravity in 11 dimensions does not match the 10 dimensions of superstring theory but this was viewed by physicists as perhaps a mathematical anomaly.

The tremendous accomplishments of superstring theory by the mid 1980's is now known as the First Superstring Revolution.   More progress soon followed.

Shortly after the development of E(8)xE(8) by Schwarz and Green, four physicists at Princeton University devised a new superstring theory.   Jeff Harvey, David Gross, Ryan Rohm and Emil Martinec, known as the Princeton String Quartet, combined features of bosonic string theory and superstrings in a new theory called heterotic string theory.   Heterotic theory postulates closed strings which are orientable.   Orientable means you can tell which direction you are moving on a string, either clockwise or counterclockwise.   That is vibrations on the string can travel in either direction.   Right moving strings in heterotic theory exist in a 10-D space-time and are supersymmetric.   Left moving strings exist in a 26-D space-time and are non-supersymmetric as in the original bosonic theory.   For this to work the 26-D space-time has to reduce to the 10-D space-time.   This was accomplished by assuming the 26-D space-time consisted of a 10-D space-time along with a 16-D curled up space-time.   Now both bosons and fermions would appear to exist in a 10-D spacetime.

In order for this theory to work it had to include either the E(8)xE(8) or SO(32) symmetry groups for the gauge forces.   These are, of course, the same gauge force symmetries required by the Schwarz and Green superstring idea that had been so useful in solving many of the earlier problems of superstrings.   So now there would be two heterotic string theories, one including E(8)xE(8) and the other utilizing SO(32), with both existing in 10 dimensions.   Since we live in a

4-D world, 6 of the 10 dimensions must be curled up.   Because superstring theory has numerous symmetries the curled up dimensions can only be compactified in a particular fashion.   An Italian-American mathematician by the name of Eugene Calabi had explored the possibility of just such compactified spaces in the 1950's.   But it wasn't until 1976 that the Chinese-American physicist Shing-Tung Yau created the formal mathematics of these restrictive curled up dimensions which are now called Calabi-Yau spaces.

The culmination of the First String Revolution is considered to be a superstring theory presented by physicists Ed Witten, Gary Horowitz, Andrew Strominger and Phillip Candelas in the mid 1980's.   Their proposal used the $E(8)xE(8)$ symmetry group along with a particular version of a Calabi-Yau compacted space.   They were able to show that at the low temperatures that we see in the universe today that the $E(8)xE(8)$ symmetry breaks down to a $E(6)xE(8)$ symmetry. The $E(6)$ symmetry group contains the $SU(5)$ symmetry group $[SU(3)xSu(2)xU(1)]$ which includes the Standard Model.   The $E(8)$ portion of $E(6)xE(8)$ includes matter not in the Standard Model. This matter would not interact with the Standard Model $E(6)$ gauge field forces (ie. EM, weak and strong) but would interact with gravity.   Astronomers have long predicted that there is missing matter because the motions of galaxies indicate that unseen or 'dark' matter is affecting their gravitational motion.   For this reason the realization that the $E(6)xE(8)$ symmetry group included matter only affected by gravity but not visible to us since the photons of our EM gauge force would not interact with $E(8)$ matter caused great excitement.   Could this be the predicted dark matter?

In addition a special version of the Calabi-Yau space manifold with three doughnut style holes would also explain the three families of matter we see.   Needless to say there was considerable excitement in the physics community that this $E(8)xE(8)$ superstring theory represented the long sought for unified theory.

Unfortunately as time went on it was realized that there were some flaws in superstring theory. For one thing it appeared there were no conclusive ways to test superstring theory. Supersymmetry is part of superstring theories but supergravity does not necessarily contain strings so even if the LHC discovers supersymmetry it will not prove superstring theory.

Other difficulties with string theory include the need to input certain constants into the theory. One of these is the string coupling constant which describes the strength of the interactions between strings.   A truly fundamental theory should tell us what this constant is.

Another problem is the particular Calabi-Yau topology that nature chooses.   Among the thousands of Calabi-Yau manifolds each can have various topologies; i.e. the holes, handles and tunnels that comprise their internal shape.   But for each topology there are many different choices for the size and shape of the compact dimensions.   These parameters are called moduli.   The moduli by affecting the vibrations of strings are what give particles their particular masses and the force coupling constants.   Unfortunately string theory does not tell us which set of moduli are used by nature to give us the world we see.

 One of the most serious failings of superstring theory is its prediction for the cosmological constant which describes the energy of the vacuum of space and is associated with the expansion of the universe.   This is the expansion of space itself which is making the universe larger. Astronomical observations indicate the cosmological constant is small but not quite zero.   String theory says if supersymmetry is unbroken then the cosmological constant is exactly zero, not quite what we are looking for.   However if supersymmetry is broken, as is believed, then string theory predicts an absolutely gigantic cosmological constant.   It is so large that it is the greatest disconnect between observation and theory known to science!   A definite problem for string theory.

Since we don't see the superpartners in our 4-D world and assuming 6 dimensions are compactified, then supersymmetry must be broken at low energies.   But superstring theory does not tell us how supersymmetry is broken.   Nor does it tell us why we see so few massless particles, i.e. the photon and gluon in our world, despite the fact that superstring theory predicts many massless particles.   The discrepancy between the 10 dimensions predicted by superstrings versus 11 dimensions predicted by supergravity also remained a puzzle.

Superstring theory despite its flaws finally culminated in five different theories.   The following table briefly summarizes these five theories:

Superstring Theories

Type	Supersymmetry	Contains Gauge Symmetry?	Oriented?	Strings Open/ Closed	Graviton	Chirality
I	N=1	SO(32)	No	Open & Closed	Yes	Yes
IIA	N=2	No	No	Closed	Yes	No
IIB	N=2	No	Yes	Closed	Yes	Yes
Heterotic-A	N=1	SO(32)	Yes	Closed	Yes	Maybe
Heterotic-B	N=1	E(8)xE(8)	Yes	Closed	Yes	Yes

All five theories contain a massless graviton so are potential candidates for a theory of gravity. Not all of them contain gauge symmetry so those that don't cannot describe our world where we clearly see the three non-gravity forces.   But all five theories exist in 10 dimensions with 6 of these dimensions compactified.

Chirality can be obtained in one of two ways.   First if the original higher dimensional 10-D superstring theory (with 9 space dimensions so mirror reflections means parity is not conserved so you have the potential for chirality) already has gauge symmetry groups and also gravity then chirality will be preserved after 6 of the dimensions are compactified.   If not then chirality can be lost as indicated in Heterotic-A and Type IIA superstring theory.   Where gauge symmetry is present in the initial 10-D superstring theory, chirality is preserved as in Type I and Heterotic-B superstring theory.

The second way chirality can be preserved is if the strings are oriented, that is vibrations on closed strings can go either to the right or left as in Type IIB theory.

Lastly all five superstring theories at low energy are effectively supergravity theories.   This strongly suggests we are unraveling the early structure of the universe at different phases of its evolution.

Of all the five superstring theories the Heterotic-B E(8)xE(8) appears to most closely resemble, at low energies with supersymmetry broken, the universe we live in.   But the big puzzle is why would nature allow five different theoretical possibilities of how the universe would be structured in its earliest moments? It would seem nature should narrow down its choices to just one.

M-Theory

It took some ten years before the next superstring revolution took place. This gives some idea of the difficulties facing the physics community in getting beyond the First Superstring Revolution.

The key to the second superstring revolution is a concept known as duality and objects called membranes. The five separate superstring theories that had resulted from the first superstring revolution were something of an embarrassment to physicists. Why would nature have five different options rather than just one consistent theory? Furthermore there was seemingly no way to test any of the five superstring theories.

Besides the problem of multiple superstring theories there was also the issue of making accurate predictions of string interactions. An equation in each of the five string theories gives the value of the coupling constant in that theory. The coupling constant equation is a measure of the strength of particle interactions within a theory. Unfortunately this equation cannot be determined for any of the five equations. The best that can be done is an approximation known as perturbation analysis.

Perturbation analysis can be useful where the string coupling constant is less than 1 or in other words where the strings interact only weakly. When strings interact strongly perturbation theory simply doesn't work.

Weakly interacting strings are only minimally affected by the frenzy of virtual string/anti-string particles that result from the uncertainty principle. However with strongly interacting strings quantum uncertainty creates so many virtual particles each of which contribute to the strength of the interaction that it becomes impossible to calculate the formula needed to estimate the string coupling constant.

Although the perturbation approach works with weakly coupling strings, physicists don't know for sure if the resulting formulas that determines the coupling constant for the various string theories is actually correct for that string theory. There is nothing to tell us whether a particular string theory's actual coupling is greater or less than 1. So by the mid 1990's it was clear that some new, non-perturbative approach would be necessary if string theory was to progress.

Fortunately the tools for just such a breakthrough were already available. The key was the fashioning of theories that went beyond one dimensional strings. Earlier in 1986 a supersymmetric theory of extended 2-D objects or membranes was created by physicists Joseph Polinski, James Hughes and Jun Liu at the University of Texas. These vibrating 'supermembranes' had the characteristics of both fermions and bosons. A year later physicists Eric Bergshoeff, Ergin Sezgin and Paul Townsend also fashioned a theory of 2-D objects in a 11-D universe with characteristics of supergravity. Paul Townsend suggested the idea of labeling these multiple dimension objects as p-branes where p would represent the number of dimensions of the object such as a 2-brane, 3-brane, etc. The maximum value of p would, of course, have to be one

dimension less than the maximum number of space dimensions in any particular string theory.   P -
branes, like strings, could behave as particles with charge, mass and spin.

Then in 1987 physicists Michael Duff, Kellogg Stelle, Paul Howe and Takeo Inami demonstrated that 2-branes wrapped tightly in a tubelike fashion resembled the vibrating 1-D strings of superstring theory.   Interestingly these 2-branes in 11-D theory looked like strings in 10-D theory!

It was around this time that the concept of dualities associated with string theories was discovered by Paul Townsend, Christopher Hull and other physicists.   Dualities in physics theories refers to the ability to exchange various parameters within a particular theory or between theories but still leave the physics predicted by the theories unchanged.   Townsend and Hull noticed that in Type III string theory that if you swapped the strongly coupled version of this theory with its weakly coupled version that the predicted physics stayed exactly the same.   This is known as strong-weak coupling duality and is labeled as S-duality.   In effect Type III is self dual. S-duality is good news for physicists since it permits understanding the strongly coupled version of Type III string theory by studying the weakly coupled version of Type III theory.   Normally the strongly coupled version of Type III would be beyond our ability to understand because perturbation theory is so complex in strongly coupled versions of string theories.

Ed Witten was inspired by these results to look at possible dualities in other string theories. Soon he discovered a similar S-duality between Type-I string theory and SO(32) string theory.

But another type of duality called T-duality was also found to be particularly interesting.   It was discovered that in certain cases a universe with radius R is identical to a universe whose radius is $1/R$ and both would have the same energy content.   Since masses and charges of particles are ultimately a function of energy then this means these two universes are indistinguishable.

Now if we allowed the universe with radius R to expand and the universe with radius $1/R$ to correspondingly shrink the energy of both universes would still be the same.   This type of duality between universes with reciprocal radiuses of their circular dimensions is called T-duality.   It turns out that Type IIA string theory is T-dual to Type IIB string theory.   Likewise SO(32) and E(8)xE(8) string theories are T-dual to each other.   Earlier we had mentioned Type I string theory is S-dual to SO(32) heterotic string theory.   Physicists in fact have been able to relate all five of the superstring to each other.

But in a physics conference in 1995 called 'Strings 95' Ed Witten astonished the string theory community by announcing an unusual result from his study of the Type IIA superstring theory. Witten had found that increasing the strength of the coupling constant in Type IIA caused the appearance of a new circular dimension making Type IIA an 11-D theory!   A short time later Witten and Petr Horava found a similar result with E(8)xE(8) superstring theory except that the extra dimension in this case was linear rather than circular.   This new 11-D theory derived from these other superstring theories can be considered to be a new theory in its own right. Interestingly Witten found that at the low energy limit the new 11-D theory was equivalent to supergravity!   The 11-D theory, however, was no longer a theory of just strings since it was found to contain 2-D membranes.

All of these string and membrane theories are connected to each other by various dualities. Furthermore all of these theories appear to be limits to another more basic and fundamental theory which is now called **M-theory**.   The exact physics of this M-theory are not known but physicists believe that through the superstring theories and the 11-D theory that they are getting glimpses of M-theory which may be the final theory of everything.

The illustration on the next page outlines what we know about M-theory from the various string and membrane theories and their interconnecting dualities.

When the E(8)xE(8) heterotic string theory is in string coupling range it not only develops a new 11th dimension (10 space dimensions + one of time) but it also reveals the existence of two separate 3-branes, each with one of the E(8) symmetries of E(8)xE(8), that act as bounds to the additional dimension.   One of the branes represents the 4-D Standard Model world we live in with an extra 6 dimensions curled up in compacted Calabi-Yau spaces.   The second brane contains the remaining E(8) symmetry particles but these are hidden from our world.   As with our Standard Model world the second brane (or braneworld) will have its own set of matter and force particles. The additional 11th dimension space separating the two braneworlds is referred to as the **bulk**.

M-Theory

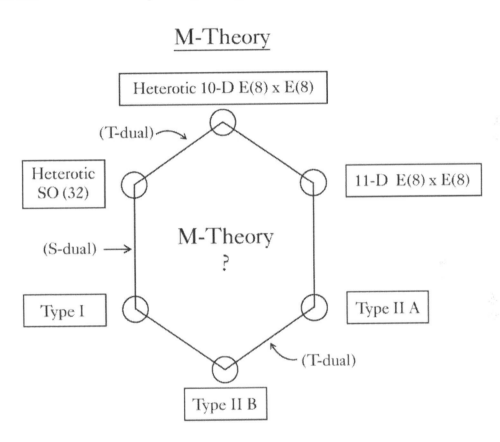

In 1989 an unusual type of brane had been discovered by the physicists Joe Polchinski, Rob Leigh, Jin Dai and Petr Horava.   It was found that open strings with their two free ends had to end somewhere.   Research indicated that these loose string ends had to attach to a type of brane known as a D-brane.   The letter D comes from Peter Dirichlet, a 19th century German mathematician, who investigated topological problems associated with the boundaries of objects. The string ends can both end on one brane or on two separate branes.   Eventually physicists concluded that D-branes and P-branes were the same thing but it was now clear that open string ends attached to branes.

In the 11-D E(8)xE(8) theory discovered by Witten and Horava it was realized that the 3-brane representing our universe was in fact a type of D-brane.   This meant that all string particles with open ends would be confined to our 3-brane universe.   Only closed loop strings like gravitons would be able to travel on the 3-brane, other branes and into the bulk.   Since the non-gravity gauge force strings are open strings in some theories this meant the electromagnetic, weak and strong force strings were confined to our 3-brane universe.   The illustration below shows this Horava-Witten braneworld:

11-D E(8) x E (8) Theory

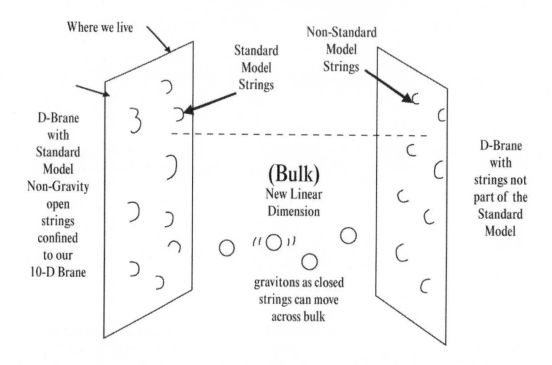

The implications of these restrictions on strings abilities to move on or off branes would lead to some amazing new theories of how our universe is constructed.   We'll take a look at one of the most interesting of these theories in the next chapter.

Branes and Extra Dimensions Strong Gravity Theories
Chapter 12 in Section II

After the Superstrings 1995 conference branes became an exciting new component of theoretical model building of our universe. Since there were still many unanswered questions in the various physics models the discovery of branes coupled with the concept of extra dimensions opened up entirely new possibilities that might lead us closer to a final theory of everything. A number of theories involving branes and extra dimensions were developed in the late 1990's.

One of the most interesting theories involving branes and extra dimensions was presented in a paper in 1999 by Lisa Randall and Raman Sundrum. Lisa Randall is a professor of physics at Harvard University. Raman Sundrum, at the time the paper was published, was a post doctoral student at Boston University but is now a professor at John Hopkins University. There were several versions of their brane ideas but for our purposes I will focus on their first model known as Randall-Sundrum 1 or simply RS1.

Randall and Sundrum were trying to solve the hierarchy problem which is in part the puzzle of why gravity appears to only be strong at around energy levels of 10^{16} TeV while the electro-weak scale unification energy is around one TeV. Roughly speaking the difference in energy scales is around 16 orders of magnitude! At the energy scales that we live in of less than one TeV, gravity is extremely weak versus the other forces of nature. Why?

The great puzzle in modern physics is why such huge energy and distance scale differences should exist in nature. Could there be some new physics that would make the hierarchy problem less of an issue? This is exactly what RS1 accomplishes. Of particular interest is that RS1 is similar to the $E(8)xE(8)$ heterotic string theory with 11 dimensions. This suggests its possible connection to the not yet fully defined M-Theory. Or perhaps RS1 is M-Theory.

The basic idea of RS1 is that the universe consists of two parallel membranes separated by and embedded in an additional 5th dimension which, as in the Horava-Witten model, is called the bulk. One of the membranes or branes is called the TeV brane and represents the spacetime we live in. The other membrane is called the Planck brane. Branes carry energy so general relativity says they will affect the spacetime around them. According to RS1 the Planck brane strongly warps the 5th dimension bulk to produce a spacetime of constant negative curvature called an anti-de Sitter spacetime. The distance between the branes is thought to be slightly greater than the Planck scale length.

The Tev brane, as its name implies, exists at the lower energy of the electro-weak scale. This brane is believed to harbor all the standard model gauge bosons and fermions as well as some gravitons. The gauge bosons and fermions are thought to be open strings with both ends attached to the TeV brane. For this reason these non-graviton particles are confined to the TeV brane. But gravitons as closed loops from string theory are free to travel on the TeV brane or bulk. However because of the intense warping of the bulk by the Planck brane most of the gravitons in the universe are confined to a "trough" in the bulk near the Planck brane. The warping of the bulk is at its greatest near the Planck brane but then becomes exponentially less warped as you move toward the TeV brane. This means a standard model particle would become heavier and smaller if it were to get closer to the Planck brane. Conversely as a standard model particle gets further away from the Planck brane and eventually gets on the TeV brane it will have the sub-TeV mass that we see in our world. A standard model particle on the TeV brane moving to the Planck brane would increase in mass by a factor of 10^{16}.   An illustration of the RS1 warped geometry model is shown below.

So RS1 neatly solves the hierarchy problem by allowing gravity to be strong near the Planck brane.   Gravity is weak in our TeV brane world simply because most of the gravitons in the universe are confined to an area close to the Planck brane.   Unification of all the forces of nature is also a possibility in RS1.   The Spanish physicist Alex Pomarol pointed out that because of the warping of the bulk, which allows the extra dimension to be small, the gauge bosons and fermions could be closed strings just like gravitons and therefore they too could travel in the bulk.   The idea is that the bulk is so small that the strength of the gauge forces would not be diluted by too many gauge bosons wandering into the bulk.   Some of the string theories do in fact call for closed strings for all the fundamental particles.   Gauge bosons in the bulk could then experience energies as high as the Planck scale and unification of the forces is then a possibility.

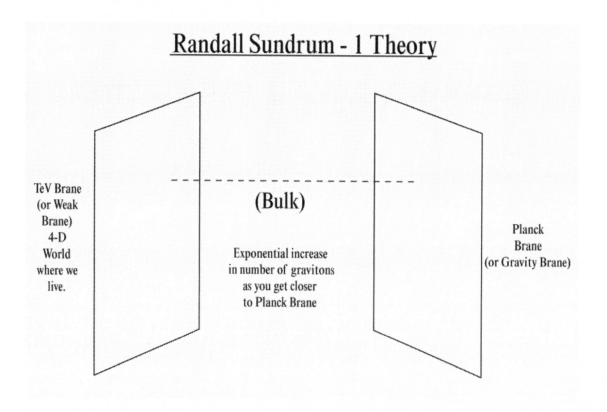

The warping of the bulk in RS1 requires the special space configuration called the anti-de Sitter spacetime.   This is a version of a de Sitter spacetime model that was developed in the early part of the 20th century by the Dutch astronomer Willem de Sitter.   He found a solution to Einstein's equations of general relativity that described a symmetrical but empty spacetime which experienced acceleration over time resulting in an ever expanding volume of space.   Einstein's original equations indicated that the universe would collapse from the effects of gravity.   Because it was believed in the early twentieth century that the universe was static Einstein added a term to his equations called the cosmological constant which counteracted the tendency of the universe to collapse.   But astronomers later discovered that the universe is actually expanding so Einstein decided to eliminate the cosmological constant.   Ironically it turns out the cosmological constant was valid anyway because quantum theory indicates that empty space does have energy from virtual particles that causes it to expand!   The de Sitter spacetime predicting an expanding universe does roughly resemble the (TeV brane) universe we see.

The anti-de Sitter version of the de Sitter equations is abbreviated as AdS.   An AdS universe has negative curvature.   A simple example will illustrate this.   If space was two dimensional then a flat piece of paper representing that space has zero curvature.   A sphere, which also has a two dimensional surface, has positive curvature.   In contrast a saddle has negative curvature.   An AdS spacetime has this type of negative curvature.

For RS1 AdS has the important characteristic of warping space such that most gravitons are confined to an area close to the postulated Planck brane.   The AdS also has the advantage that it solves the problem of the theoretical value of the cosmological constant being too large.   The best current theories in physics give a value for the cosmological constant of around 10^{120}.   This is way too large!   Astronomical data indicate that the constant should be small or possibly zero.   If it weren't small the universe would have blown itself apart long ago.   Yet the best theories give an estimate that is so out of wack that it is considered to be the largest discrepancy between theory and observation in all of science!   But in RS1 the anti-de Sitter spacetime has a negative cosmological constant which can be adjusted to offset the large positive cosmological constant derived from quantum cosmology.   Since the actual measured cosmological constant is very small the AdS allows physicists to adjust its warping to just the right amount such that the large positive brane cosmological constant minus the adjusted AdS constant equals the best observational estimate of the cosmological constant.

The fact that AdS solves the cosmological constant problem is very important.   It lends significant credibility to the RS1 theory as a strong candidate for describing our universe.   But there is more!

One of the most exciting outcomes from RS1 is the possibility of creating very strong local gravity fields.   Prior to the extra dimension theories it was assumed gravity only became strong at extremely tiny distances around the Planck length scale.   Planck scale energies are far beyond the capacity of any conceivable particle accelerator.   But if gravity is weak only because most gravitons are in the bulk between the branes then there is the real potential of creating strong gravity at much lower energy levels.   RS1 predicts that some Kaluza-Klein (KK) gravitons produced in the Large Hadron Collider will have gravitational interaction strengths sixteen orders of magnitude greater than normal gravitons.   These strongly interacting KK gravitons are partners of the ordinary 4-D gravitons.   Besides the vastly greater interaction strength of the KK gravitons some will also have mass unlike the massless 4-D graviton.   The mass/energy range of these KK graviton partners should be a tower of particles of about 1 TeV, 2 TeV, 3 TeV, etc.   Human created strong gravity fields will certainly be useful in teaching us much more about nature at its deepest levels.   The implications of manmade strong gravity fields and their possible association with the UFO phenomenon will be of particular interest.

In the next and last section of the book, Section III, I'll summarize where we are in modern physics and also do a brief review of the most important UFO characteristics.   With this background I will then propose a solution to the UFO enigma based on merging our latest understanding of physics with pertinent UFO data and some conjectures I have developed in recent years.   As we shall see one of the keys to solving the puzzle of UFO technology is the extra dimensions and brane theories.

Section III

Solving the UFO Enigma

Physics Summary
Chapter 1 in Section III

Classical Physics Before the 20th Century

Classical physics includes all physics before the development of quantum mechanics in the 1920's. For this reason relativistic physics, which was formulated between 1905 and 1915, is also considered classical.   However due to its importance I will treat relativistic physics separately and use the 20th century as a convenient dividing line for pre and post 20th century classical physics. Of classical physics before the 20th century Newtonian mechanics and Maxwell's theory of electromagnetism are probably the most important and better known contributions to pre-modern physics.

Newton's laws of motion describe how objects move and interact.   His theory of gravity held sway for more than two centuries until improved upon by Einstein's general theory of relativity. For most situations Newton's laws are adequate for describing mechanical systems including the orbits of the planets.

Maxwell's laws of electromagnetism united the electric force and magnetic force, once thought to be separate forces, into a single force we call electromagnetism.   He published his equations of electromagnetism in a 1864 paper titled "A Dynamical Theory of the Electromagnetic Field." Maxwell's laws required that all observers must measure the speed of light to be the same but this went unnoticed for some four decades.

An important consequence of Maxwell's uniting of the electric and magnetic forces into a single electromagnetic force was the introduction of the local gauge symmetry concept.   For example physicists found that changing the electric field charges produces a compensating magnetic field such that the combined electromagnetic field stays invariant.   A local gauge particle/field, the photon, is created to transmit information about the changed electric field potential and instigates the compensating magnetic field potential.   This concept has been very helpful in understanding the forces of nature and predicting local gauge quantum particles involved with compensating force fields such as the W+, W- and Z neutral weak force gauge particles of the combined electro-weak force.

Relativistic Physics

Einstein's two best known contributions to modern physics were the special theory of relativity and the general theory of relativity developed in the early 20th century.   These equations modify Newton's laws of motion in extreme situations such as objects moving at near light speeds or masses subject to a strong gravity field.

Special relativity showed that the constancy of the speed of light to all inertial observers, i.e. non-accelerated motion, meant that space and time were not independent entities.   Taken together space and time are invariant but relative to each other both space and time can vary.   The notion that space and time are distorted for objects traveling at close to the speed of light had already been suggested in the Lorentz-Fitzgerald transformation equation which was incorporated into special relativity.   A consequence of the malleability of space and time is Einstein's famous equation equating mass and energy or $E = mc^2$.

Minkowski later showed that time could be treated as a fourth "spatial" dimension by using the speed of light as a conversion factor.   This 4-D geometrical version of Einstein's special relativity is where we get our modern version of space-time.

General relativity had its inspiration from Einstein's realization that a person falling from a roof does not feel any gravity force.   The sensation of weight is exactly canceled out by the acceleration of their fall.   This means the effects of acceleration are exactly equivalent to the effects of a uniform gravitational field.   This relationship between acceleration and gravity results from Einstein's equivalence principle postulate that states that gravitational mass and inertial mass are the same.   The classic example of this is to imagine someone in a rocket ship out in deep space accelerating at 1 G.   The individual in the rocket ship will be pressed against the floor of the rocket with a force of 1 G and will be unable to distinguish between the artificial gravity from the accelerating rocket ship and the gravity experienced on the surface of the earth.

With the equivalence of acceleration and gravity in mind Einstein next imagined a glass elevator in free fall dropping toward the surface of the earth.   Since the occupants of the elevator would feel no forces they would expect, according to Newtonian physics, that a light beam shot from half way up the wall on one side would hit the wall on the other side exactly half way up. This is exactly what happens.   (If they had been in deep space far away from any gravity fields and not accelerating they would see exactly the same result too).   People on the ground watching the elevator fall to the earth would also see the light beam start half way up one wall and hit the opposite side half way up that wall.   However the ground based observers would also notice that while the light beam traveled between the walls the elevator dropped closer to earth.   This means from their point of view the light beam curved downward in order to hit the opposite wall half way up.

With this simple gedanken (thought) experiment Einstein demonstrated that space-time is *curved* in a gravitational field.   For light beams and everything else the shortest distance between two points is the geodesic defined by the curvature of space-time.

The curvature of space-time in a gravity field is usually illustrated by imagining a stretched 2-D horizontal rubber sheet with a bowling ball plopped into the middle.   The bowling ball sinks into the center of the sheet stretching and distorting it at that distance.   The analogy is imperfect since it does not include time and is only two dimensions but it does illustrate the essential concept of the geometric warping of space-time.

One of the interesting consequences of general relativity is that time slows down in a gravitational field.   The slight difference in the passage of time between a clock on the earth's surface and a spacecraft deep in space has actually been measured and conforms to the predictions of general relativity.   Other predictions such as the bending of light beams have been observed by astronomers in a phenomenon known as gravitational lensing.

Despite the fact that general relativity is based on the equivalence of inertial mass and gravitational mass Einstein was unable to explain the origin of inertia in his theory.   Since UFO's appear to defy the laws of inertia this inability to explain it is of particular interest.   We'll take a closer look at inertia next.

Inertia

Inertia is defined as the resistance of a body to acceleration.   The greater the mass of an object the greater its resistance to acceleration.   For example imagine the force necessary to accelerate from rest a marble versus a bowling ball.   Note that in applying force to a stationary object it causes the object to go from zero speed to some higher speed.   This change in speed is acceleration which is produced by a force.   Newton determined the math relationship for the force needed to accelerate a mass.   His second law of motion is:

F=ma   **Where F=force**
                  m=mass
                  a=acceleration

Associated with inertia is the concept of ***inertial mass*** which is a measure of the amount of matter in a body based on its resistance to acceleration.   Inertial mass is determined from the above formula to be:

m=F/a

You can also measure the mass of an object based on its gravitational interactions.   This is known as an object's ***gravitational mass***.   The force of gravity between two objects is determined by Newton's gravity law:

$$F = G \cdot \frac{m_1 m_2}{r^2}$$ **Where: F = gravity force between objects**

G = **universal gravitational constant**

m_1 **and** m_2 **are the masses of the**
two objects.

r = distance between the two masses

The intriguing thing about gravitational mass and inertial mass is that they are identical based on numerous experiments.   It is not clear why this should be so.   As John Gribbon in his book "Schrodinger's Kittens and the Search for Reality" puts it: "Gravitational mass determines the strength of the force which an object extends out into the Universe to tug on other objects; inertial mass, as it is called, determines the strength of the response of an object to being pushed and pulled by outside forces – not just gravity, but *any* outside forces.   And they are the same.   The 'amount of matter' in an object determines both its influence on the outside world, and its response to the outside world.   This already looks like a feedback at work, a two-way process linking each object in the Universe at large."

The Austrian physicist Ernst Mach proposed that the inertia of an object is due to all the other masses of the universe which tug gravitationally on the object from all directions so any attempt to move the object is resisted by the gravitational pull of the rest of the universe.   Einstein was impressed with this concept which he labeled **Mach's principle** and tried to incorporate it into his general theory of relativity.   But unfortunately in order for Mach's principle to be used to explain inertia in general relativity it would require a feedback mechanism that works *instantaneously*. Special relativity forbids any signals traveling faster than the speed of light so there is apparently no possibility of instantaneous gravitational feedback from distant parts of the universe.

A number of ideas have been proposed to explain inertia but so far none have been widely accepted or supported by any experimental evidence.   One interesting idea is an extension of the Wheeler-Feynman absorber theory.   This idea was originally developed to explain something called radiation resistance.   When you try to accelerate a charged particle like an electron it reacts with more resistance than just inertial resistance from its mass.   An additional resistance above and beyond its inertia also appears.   It is thought that this is due to the charged particle reacting to all the other charged particles in the universe!   But how can this be since again these other charged particles in the universe can be millions of light years away and no signal can travel faster than the speed of light?   The Wheeler-Feynman absorber theory explains radiation resistance by assuming that electromagnetic waves can travel both forward and backward in time.   While we won't go into details on this theory, the concept allows for instantaneous reaction of an accelerating charged particle to all the other charged particles in the universe.

The Wheeler-Feynman idea can be extended to inertia as well if you assume that virtual gravitons also travel both forward and backward in time.   This was done by Professor John Cramer in the 1980's at the University of Washington.   Cramer pointed out that the relativistic version of the Schrodinger equation, which is used as the basis of quantum mechanics, has two solutions.   One solution describes quantum waves going forward in time while the other describes quantum waves going backward in time.   Normally the backward in time quantum waves are ignored by physicists under the assumption that they make no sense.   But Cramer proposed that the solution allowing waves to travel backward in time should be considered a viable solution of the Schrodinger equation.   If so then this could explain, via an extension of the Wheeler-Feynman absorber theory, how the inertia of a body results from the instantaneous feedback of all the matter in the universe.   This is called the **transactional interpretation** of quantum mechanics.

Whether or not Professor Cramer's transactional interpretation is correct is not known.   But the key idea to be understood in this entire discussion on inertia is the instantaneous nature of the inertial force and the need to explain it.   As noted previously in the summary of general relativity an understanding of inertia will almost certainly be involved in an explanation of the UFO phenomenon.   More on this later.

Next we'll do a brief review of quantum mechanics.

Quantum Mechanics

Quantum mechanics explains the physics of the subatomic world which classical physics could not describe.   But classical physics itself is a special case of quantum physics where large numbers of quantum entities (atoms, molecules) are involved such as in a cannon ball. When dealing with large scale object's Planck's constant becomes irrelevant.

Quantum mechanics is built on some fundamental equations, principles and observed quantum behaviors. We'll just briefly review some of these key constituents of quantum theory.

Wave Mechanics and Matrix Mechanics

Wave mechanics, represented by Schrodinger's equation, is one of two ways to explain the behavior of quantum particles like electrons. The other approach is called matrix mechanics developed by Werner Heisenberg. Both theories were developed in the mid 1920's. Wave mechanics, as its name implies, treats quantum entities as waves. Matrix mechanics treats quantum entities as particles. This reflects the wave-particle duality seen in other areas such as whether light consists of waves or particles.

Schrodinger's equation was inspired by DeBroglie's suggestion that if electrons were waves then only a whole number of oscillations of the wave would fit in an orbit around the atomic nucleus. At first Schrodinger could not get his equation to match the energy levels of atoms calculated from spectroscopic analysis. But once he took into account relativistic effects he got a close match to spectroscopic data. It was later shown by Dirac that the relativistic effect is associated with the spin of the electron. Of course an interesting question is how does a wave spin?

As we saw in the chapter on quantum mechanics the Schrodinger equation gives us the wave function of a quantum entity such as an electron. Max Born developed a probabilistic interpretation of Schrodinger's wave equation which requires taking the absolute value of the square of the wave function. The absolute value is needed because probabilities can only be positive, i.e. 0% to 100%.

So the Born interpretation of Schrodinger's equation gives us only the probability that a quantum entity like an electron will be found in a particular location. This probabilistic rather than deterministic view of nature prompted Einstein to retort that God does not play dice with the universe.

Although wave mechanics does correctly predict quantum phenomena it's important to note that we don't know if quantum particles are really waves.

Matrix mechanics developed in 1925 by Heisenberg was the first self-consistent explanation of quantum mechanics. Heisenberg noticed that all attempts to explain quantum processes used classical physics analogies such as orbits and spin despite the fact that the actual behavior of subatomic particles could not be observed. He realized that the only thing we really know about quantum activity is the beginning and ending state of a quantum process. For this reason Heisenberg decided to just focus on the mathematics of pairs of quantum states, the start and end state, such as when an electron is thought to jump from one "orbit" to another.

In using pairs of quantum states Heisenberg soon realized he could not use ordinary mathematics to describe the quantum world. Instead he used matrixes which are arrays of numbers arranged in squares like a checkerboard. Unlike ordinary numbers Heisenberg also discovered that these matrixes were mathematically rather unusual. He found for example that multiplying matrix A times matrix B did not equal B times A. Using his matrix concept Heisenberg was able to replicate the equations of Newtonian mechanics including important laws such as conservation of energy.

A formal version of Heisenberg's discovery was fully developed by Heisenberg along with Max Born and Pascal Jordan.   Born made the key insight that Heisenberg's unusual math was actually a somewhat obscure but well developed branch of mathematics known as matrices.   The so called 'three-man paper' explaining matrix mechanics was published in 1925.

Quantum mechanics, like general relativity with gravity, has been highly successful in describing the subatomic world.   Yet there are puzzling quantum phenomena which can be predicted but nevertheless raise other questions.   We'll look at two of these phenomena, non-locality and quantum tunneling, next.

Non-Locality

Non-locality refers to how quantum particles can be affected not only by local interactions but by non-local events that in principle could occur on the other side of the universe.   But the puzzling feature of quantum non-locality is that, like inertia, it occurs *instantaneously*.   Special relativity says nothing can go faster than the speed of light yet here is a phenomenon that clearly violates that rule.   It is no wonder that Einstein referred to non-locality as 'spooky action at a distance'.

Experiments have demonstrated that non-locality is a undeniable feature of our world.   One of the better known experiments proving non-locality was conducted by the French scientist Alain Aspect in the early 1980's.   His experiment used pairs of photons emitted simultaneously in opposite directions from an excited atom.   Polarization detectors that could be set at various polarization angles were set up on opposite sides of the paired photon emitter source.   The exact details of the experiment are not important but the resulting pattern of photon detections matched quantum mechanics predictions.   The implication is that somehow information is passed between the photons instantaneously.   This contradiction with special relativity cannot currently be explained.

Tunneling

Another interesting quantum effect which allows subatomic particles to penetrate seemingly impossible barriers is a process called tunneling.   The uncertainty principle tells us that if you know the momentum of a particle to a high degree of accuracy then the position of the particle becomes highly uncertain.   Nature makes use of the uncertainty principle in the production of energy via nuclear fusion.   When two protons come close together their positive electric charge causes a repulsion between them.   But in the interior of a star the high momentum of protons allows them to come near each other.   The uncertainty principle then dictates that there is a certain probability that sometimes a proton will appear on the other side of the electric barrier between the protons.   When this happens fusion takes place and energy is released.

Radioactive decay is another example of tunneling.   Again the uncertainty principle means that occasionally a proton in a nucleus will find itself outside the range of the strong force that binds the nucleus together.   This is called alpha decay.

Like non-locality, quantum tunneling has been found to occur faster than the speed of light.   Again this violates special relativity but there is as yet no explanation for this puzzling quantum behavior.

Standard Model

The Standard Model combines quantum mechanics with special relativity for all the fundamental particles and forces of nature except for gravity. With the exception of gravity the Standard Model does a superb job explaining nature at its deepest levels.

The Standard Model says that all matter particles are either leptons or quarks or some composite of these basic particles. These fundamental particles interact via the four known forces of electromagnetism, the weak nuclear force, the strong nuclear force and gravity. The interactions of matter particles via these forces, with the exception of gravity, are now well understood at a quantum level. In the Standard Model chapter we saw that QED, or quantum electrodynamics, explains to a high degree of accuracy the exchange of photons between charged particles in the electromagnetic interaction. QCD, or quantum chromodynamics, describes the interaction of quarks via gluons in the strong interaction. QED was later expanded to include the weak interaction combined with the electromagnetic interaction in the electroweak theory.

A key element in the successful completion of the electroweak symmetry theory was the proposed Higgs particle(s). The Higgs bosons are believed to have become active only after the early universe cooled to a certain temperature. At that point the electroweak symmetry was broken into the weak force and electromagnetic force after three of the four electroweak force carrier particles gained mass via the Higgs mechanism.

Various unsuccessful attempts have been made to unify the electro-weak and strong force interactions in a GUT or grand unified theory. But nevertheless QED, electro-weak and QCD are highly successful theories that have given us an extraordinary understanding of nature.

Only gravity has so far eluded efforts to understand it at a quantum level. General relativity does explain gravity as the geometric warping of space-time but unfortunately not in terms of the postulated quantum entity we call the graviton. A theory of quantum gravity remains as a kind of Holy Grail for modern physics and recent theories hold some promise of reaching this goal. Gravity is incredibly weak compared to the other three forces of nature and this is one of the major reasons that a theory of quantum gravity eludes us. It should not be too surprising that gravity, the one area of physics that is outside the Standard Model and not completely understood by physicists, appears also to be at the heart of the UFO enigma.

Beyond the Standard Model

We'll do just a brief review of the major theories that go beyond the Standard Model.

Grand Unified Theories

The Grand Unified Theories or GUT's theoretically unite the electromagnetic, weak and strong forces at high energies. The coupling constants that describe the strength of the three non-gravity force interactions are far apart at low energies. But according to several GUT theories these coupling constants move very close together at around 10^{15} GeV. One of the most promising early GUT theories used the SU(5) symmetry group.   Unfortunately the coupling constants came close in the GUT's but were not exactly the same.   Close but no cigar.   There were other problems too such as the prediction of proton decay which has so far not been seen.

4-D Supersymmetry

Supersymmetry theory completes the last significant symmetry that may exist in nature. This last symmetry is the distinction between fermions and bosons.   Supersymmetry postulates an entire new set of matter and force particles that are partners to the known matter and force particles. The partnership matches known fermions to supersymmetry bosons and known bosons to supersymmetry fermions.   The end result is the doubling of the total particles in our 4-D world. So far no evidence of supersymmetry has been found but physicists hope the Large Hadron Collider will find at least the lightest of the superpartners.

4-D Supergravity

Supergravity is an extension of 4-D supersymmetry and predicts the existence of a spin-2 graviton along with its superpartner the gravitino with spin 3/2.   What is exciting about supergravity is that it is a quantum theory of gravity which means that it predicts the existence of a quantum particle, the graviton, that carries the force of gravity.   Note that this first version of supergravity is based on the standard four dimensions.

Eventually a total of eight supergravity theories were devised with ever increasing numbers of particles including fermions and bosons.   The eight different supergravity theories are labeled N=1, N=2, N=3, etc. up to N=8.   The N number refers to the number of gravitinos in each theory, for example N=3 is the supergravity model with three gravitinos.

Unfortunately 4-D supergravity has serious problems that tell us it cannot represent nature as we know it.   The most promising supergravity group of O(8) does not include the full Standard Model symmetry of SU(3)xSU(2)xU(1).   In addition chirality is missing and the gravitational infinities are not completely vanquished.

5-D Kaluza-Klein and 11-D Supergravity

In 1919 Theodor Kaluza modified Einstein's theory of general relativity by adding one more space dimension to make it a 5-D theory.   He discovered that both Maxwell's 4-D equations of electromagnetism as well as general relativity could be derived from this 5-D version of Einstein's theory.   In 1919 the only forces known were gravity and electromagnetism so Kaluza's unified theory created great excitement as a possible theory of everything.

In 1926 Oscar Klein modified Kaluza's idea to take into account quantum theory which was developed after Kaluza's 5-D theory.   Klein proposed that there is really only one force, the gravity force, which in its wave configuration vibrates in a tiny curled up 5th dimension to create the force we call electromagnetism.   Although Kaluza-Klein theory stagnated for several decades due to various problems it was not completely forgotten.

In the late 1970's and early 1980's interest in Kaluza-Klein type theories with extra dimensions was revived.   It was found that by adding a total of 7 additional dimensions to the known 4 dimensions for a total of 11 dimensions that it is possible to explain all of the four fundamental forces in nature.   Since the 3 non-gravity forces had already been unified to some degree, the new

11-D theory that also encompassed gravity caused great excitement.   The theory is known as 11-D supergravity.

But as work on the theory proceeded certain problems arose such as the absence of chirality that raised doubts on the completeness of the theory.   11-D supergravity does appear to represent the universe at some earlier higher energy phase.   So despite the problems progress had been made and it looked like physicists were heading in the right direction.

10-D & 26-D String and Superstring Theory

11-D supergravity assumes that all the fundamental particles are zero dimensional points. But in the late 1960's it was discovered from experiments at CERN that hadrons behaved like tiny wiggling strings.   String theory was born!   Theorists soon concluded that there were two types of strings; open and closed.   The vibrations of these strings correspond to different fundamental particles.

String theory started out describing matter particles but it was soon extended to force particles too.   The theory only made sense with more than 4 dimensions which made it the first theory to require additional dimensions.   String theory required a 26-D spacetime for bosons and a 10-D spacetime for fermions.   In time it was found that string theory predicted the existence of the spin-2 particle we call the graviton.   While string theory resolved a number of problems that plagued earlier non-string theories it also had problems of its own such as not being able to explain the masses of the fundamental particles.

In 1976 it was discovered that a string theory designed for fermions called spinning string theory could be made supersymmetric with equal numbers of fermions and bosons.   A formal version of strings with supersymmetry was soon developed and named superstring theory.   Just as with supergravity there are eight versions of superstring theory for N=1, N=2, N=3, etc. up to N=8. The N number refers to the number of gravitinos in each version of the theory.   N=1 was found to be what is called Type I theory or a theory that can have open and/or closed strings.   N=2 is a Type II theory with closed strings.   And so on.

Like the original string theory superstring theory only makes sense with extra dimensions.   It requires 9 space dimensions and one time dimension for a total of 10 dimensions.   The original 26-D bosonic spacetime is assumed to reduce to a 10-D spacetime by having 16 dimensions curl up into a tiny spatial configuration.   Then both bosons and fermions would appear to exist in 10 dimensions.   At this point the effectively 10-D bosons and 10-D fermions would each have 6 of their 10 dimensions compactified into tiny configurations called Calabi-Yau spaces.   What is left is the 4-D world we see.   The exact Calabi-Yau configuration has not been determined.

A promising version of superstring uses the symmetry E(8)xE(8).   Of particular interest is that at the low temperature of our current universe the E(8)xE(8) symmetry breaks down to E(8)xE(6). The E(6) group contains the SU(5) symmetry group which contains the Standard Model symmetry group of [SU(3)xSU(2)xU(1)]!

The E(8) portion of the E(8)xE(6) symmetry group is the equivalent of another universe which includes matter that is not in the Standard Model.   This matter would not interact with the E(6) gauge fields but would have a gravitational interaction so may be the source of the dark matter predicted by astronomers.

Eventually superstring theory culminated in five different versions, all existing in 10 dimensions.   The Heterotic B theory with symmetry of E(8)xE(8) seems to most clearly resemble our world at low energies.   But physicists puzzled over why there should be five versions of superstring theory instead of just one final theory.

The five superstring theories are shown below.   Note that most of the theories actually have closed strings for all fundamental particles, not just the graviton.   This is important for unification of the forces as explained in Section II for Randall-Sundrum-1.

Superstring Theories
(All in 10 Dimensions)

Type	Supersymmetry	Contains Gauge Symmetry?	Strings Open/ Closed	Chirality
I	N=1	SO(32)	Open & Closed	Yes
IIA	N=2	No	Closed	No
IIB	N=2	No	Closed	Yes
Heterotic A	N=1	SO(32)	Closed	Maybe
Heterotic B	N=1	E(8)xE(8)	Closed	Yes

In time more problems were found with superstring theory that began to cause concerns.   The biggest problem is superstrings prediction that the cosmological constant is huge when it is known from observation that the constant is small.   Another concern is why does superstring theory predict 10 dimensions while supergravity predicts 11 dimensions?   But despite these problems physicists were encouraged because some progress had been made and it seemed as if they were getting closer to the ultimate truth.

M-Theory

In the 1980's a new type of object was discovered in supersymmetry theory that went beyond one dimensional strings.   These objects, called membranes, could have 2 or 3 or 4 etc. dimensions. Later they were labeled p-branes with p representing the dimensionality of the object.

In 1987 several physicists demonstrated that 2-branes wrapped tightly in a tube like fashion resembled 1-D strings in 10-D superstring theory.   In other words 10-D superstring theory has a hidden dimension so in reality it is an 11-D theory!   Just what physicists were looking for.

At about the same time that p-branes were introduced a new concept called dualities was developed by a number of physicists.   Dualities allow physicists to exchange various parameters between similar theories but still leave the physics predicted by the theories unchanged.   Using this concept it was soon determined that some of the five superstring theories were dual with each other.   In effect they weren't separate theories but related to each other.

In the mid 1990's the duality concept led to the development of a new version of the E(8)xE(8) superstring but with 11 dimensions!   This new version included 2-D membranes so it could no longer be considered just a string theory.   The discovery of branes opened up whole new possibilities in understanding our universe.

Now there were six theories that had morphed out of the original string theory but not all were based exclusively on strings.   However it was noticed that all six theories appear to be limits of another more fundamental theory which is now called M-theory.   This undeciphered M-theory could be the theory of everything.

Randall-Sundrum 1

One of the most intriguing theories to use the brane concept was developed in 1999 by Lisa Randall and Raman Sundrum.   Their theory is referred to as RS1.   The basic idea of this theory is that the universe consists of two parallel membranes embedded in and separated by a 5th dimension which in brane theory is called the bulk.   The distance between the branes is thought to be somewhat larger than the Planck scale.

One of the branes is called the TeV brane and represents the 4-D space-time universe we live in.   The TeV brane harbors all the Standard Model force and matter particles.   As the name implies the TeV brane is at the lower energy of the electro-weak scale.   The second parallel brane, called the Planck brane, is an entire other universe with its own possible set of matter and force particles which do not interact directly with particles on our TeV brane.   According to RS1 the branes carry energy so via general relativity they can affect the space around them.   The Planck brane is theorized to strongly warp the 5th dimension bulk to produce a space-time of constant negative curvature.   Gravitons as closed loops in M-theory are able to travel anywhere in the bulk or on either brane.   But because of the intense warping of the bulk by the Planck brane this causes most of the gravitons in the RS1 universe to be confined to an area near the Planck brane.   This is why gravity is so weak on the TeV brane, our universe, because the probability of finding gravitons on our brane is very low.

In some string theories even the Standard Model particles are closed loops like the graviton and can travel in the bulk.   But because the bulk is so small not many Standard Model particles will wander into the bulk.   Unlike the graviton the probability of finding Standard Model particles in the bulk is small.

The warping of the bulk is greatest near the Planck brane but becomes exponentially less warped as you get away from the Planck brane until reaching the TeV brane where energy levels come down to the sub-TeV level of our universe.   This means that any Standard Model particles that do go into the bulk will become exponentially heavier and smaller as they get closer to the Planck brane.

RS1 is of great interest because it provides a solution to the hierarchy problem and other issues. I believe RS1 or something like it will also be involved in helping to solve the UFO enigma. We'll examine this idea shortly.

UFO Characteristics Summary
Chapter 2 in Section III

UFOs would be quite unremarkable were it not for their extraordinary performance characteristics, unusual shapes, frequently reported intense radiation, often inexplicable behavior and other puzzling attributes.   In fact it is the sum total of these UFO characteristics that make them such an enigma to our human societies.   From a technical point of view I believe that by systematically categorizing UFO characteristics we can get some insight into the engineering used in these machines.   There is good reason to believe that they are machines of some type since numerous well qualified witnesses in close encounters have reported mechanical landing struts, metallic skin surfaces and transparent glass-like domes on top.   We will then use this compilation of UFO characteristics in the next chapter (Chapter 3 in Section III) along with current theories in physics to help us come to a proposed solution of the technology behind the UFO phenomenon.

UFO Configurations

One of the most interesting features of UFOs is that they almost always have a circular shape in some part of their structure.   The typical UFO as we saw in Section I is often described as a flying saucer or as a frying pan without the handle.   Other shapes include cylinders, spheres or cones. In each case the circular aspect is obvious..   Even where UFOs have other shapes such as the large flying triangles seen in the Hudson Valley region of New York State in the 1980's and over Belgium in the 1980's and 1990's, it was reported that at each corner of the triangle there were large glowing white circles.   I think it is safe to say that this circular aspect of UFOs probably has something to do with the "engine" that powers these craft.

Rapid Acceleration/Deceleration

Another key characteristic of UFOs is their often rapid accelerations and decelerations which seem to defy inertial forces.   Pan American pilots Nash and Fortenberry in their 1952 sighting, described in Section I, observed sudden changes in direction that vastly exceeded the capabilities of any human engineered aircraft.   My wife and I some years ago observed a UFO dropping suddenly from a stationary position at around 2,000 feet all the way to tree top level in just a second or so.   The Belgian Air Force in the 1980's had an instance where a F-16 fighter jet locked onto a UFO but then observed the object drop 3,000 feet to near ground level in 3 seconds or the equivalent of 700 miles per hour!   This interesting UFO encounter was recorded on the F-16's onboard radar system.   High ranking Belgian NATO officials held a press conference after the event in which they displayed these radar tapes of the unknown's phenomenal flight characteristics.

Another fascinating case involving inertia defying accelerations and rapid changes in direction was brought to the attention of Dr. J. Allen Hynek by the widow of a former U.S. Air Force navigator.   This sighting is also included in Section I.   The incredible maneuverability of the UFO in this incident was so spectacular to the pilot and crews of three B-47 bombers that they doubted anyone would believe them.   The details of this 1954 sighting were kept secret until the navigator's widow revealed this story to Dr. Hynek years later. The missile shaped UFO, similar in size to the fuselages of the B-47's, circled around the bombers at such high speeds that at times it was just a blur. At one point the UFO performed a perfect figure eight directly in front of the bombers.

There are countless other stories of UFOs exhibiting extreme maneuverability that would destroy any conventional earth built aircraft along with its human occupants. Somehow UFOs seem to be able to overcome inertial forces.

Falling Leaf Motion, Rotation and the Estimated Weight of UFO's

Another feature of UFOs is their often reported falling leaf or fluttering motion. There are numerous examples in Section I of this wobbling motion of UFOs which seems to occur primarily when these craft are moving at a slower speed. An excellent example of this type is a 1950 Radar-Visual reported by six U.S. Navy fighter-bomber aircrew during the Korean War. The Navy aircrews observed two huge circular UFOs racing toward their flight at around 1,000 to 1,200 MPH. More than a mile from the Navy formation the objects suddenly stopped in mid air. The 600 to 700 foot diameter craft were then seen to "jitter" or "fibrillate" in their almost stationary position. The full report can be seen in Section I.

Other observers describe this UFO motion as a "rocking motion", "oscillation", "fluttering", etc. This type of motion is what you would expect of a light weight object like a leaf or kite as it moves through the atmosphere. Yet there is circumstantial evidence that UFOs are massive devices of considerable weight. Ted Phillips, Director of the Center for Physical Trace Evidence, has completed studies involving soil compression at UFO landing sites which suggest that even the smaller craft range in weight from 8 to 18 tons. In a 1954 landing on railroad tracks at Quarouble, France engineers examined indentations in the railroad ties and estimated that the medium sized UFO that was seen had weighed 30 tons.

Clearly the UFO's seem to eliminate the effect of earth's gravity on their mass causing these craft to behave as if they are nearly weightless. Almost certainly associated with the UFOs ability to reduce its apparent weight is the frequently reported observation that these objects spin or rotate on their vertical axis. As anyone who has ever thrown a Frisbee knows, you can stabilize a light weight object simply by setting it into a spinning motion. In Section I there are numerous examples where the UFO itself or at least the outer rim is seen to be rotating.

Luminosity

One of the more common characteristics of UFOs is the intense radiation they are seen to generate in the visible spectrum. This self luminosity is usually seen at night but sometimes in daylight as well. A frequent description is that the craft seemed "red hot", like a glowing stove top. The changing colors and intensity appear to be related to the craft's acceleration, deceleration or other maneuvers.

The former German mayor's 1952 sighting may be significant in relating the changing luminosity of UFOs to the craft's propulsion system. In that sighting the witnesses observed holes open along the edge of the craft from which appeared glittering green light, later turning to red.

This occurred as the propulsion system of the UFO was apparently just being turned on and as it rose off the ground.   This sighting can be found in the Classic UFO Cases in Section I.

Another case that may shed some light (no pun intended!) on the relationship of UFO's luminosity to their propulsion system is the 1952 NL (nocturnal lights) sighting by two commercial airline pilots, Nash and Fortenberry, mentioned above.   In this case the crew of the Pan American DC-4 flight near Norfolk, Virginia observed six discs glowing a brilliant red-orange color, like red hot metal, racing toward them but at a lower altitude.   The discs were flying in the flat position and were in echelon formation.   The leading disc suddenly slowed and flipped up on edge and then the five trailing discs quickly executed the exact same maneuver.   The leading disc then flipped back to the flat position, shooting off in a completely different direction.   The other five discs almost instantly executed the same maneuver as their leader, falling back into the same lineup as before with the whole formation accelerating together on a new course.

As the pilots watched this display in awe, two more discs raced out from under the DC-4 and accelerated to catch up with the other six discs.   As they accelerated the two discs brightness increased noticeably.   Previously the other six discs had dimmed as they slowed down before making their turn.   But after making the turns the original six discs had also brightened as they accelerated up to "cruising" speed.

Almost certainly the luminosity and the changes in color and intensity of UFOs is somehow related to their propulsion systems or possibly is ionized air that is involved with shielding the craft during high speed flight in the atmosphere.   Of particular interest is that the luminosity is often concentrated at the rim area of circular UFOs.   This last characteristic strongly suggests that the "engine" powering these craft is located in the rim area of at least the circular devices.

Gravitational And/Or Suction Effects

A puzzling feature of many close encounters is the repeated contention of witnesses that they felt "weightless" or a sense of being pulled upward toward the UFO.   Others describe a "vacuum cleaner" sound and seeing leaves, dust and other objects being lifted off the ground beneath UFOs.

In many cases where UFOs are seen hovering over bodies of water witnesses report the water surface "mounding" under the craft.   Police officers observed this effect in the famous Wanaque Reservoir sightings in northern New Jersey in 1966.   Sergeant Ben Thompson during one sighting saw that the water's surface under the UFO was raised up 2 to 3 feet above its normal level over an area several hundred feet long.   The UFO was described as a brilliantly lit disc shaped object. UFO researcher Carl Feindt maintains a website at waterufo.net where he lists numerous other cases of UFOs pulling up water toward their bottom surface when they have been observed over ponds, lakes or the ocean.

An important UFO close encounter occurred in the spring of 1968 that Dr. Hynek suggested "may point to the physics of the UFO."   I believe his scientific instincts were right.   This case occurred in Cochrane, Wisconsin in 1968 and involved two primary witnesses, a schoolteacher and a former U.S. Air Force stewardess.   While driving on a lonely country road they encountered an orangish-red object that caused their engine, headlights and radio to fail.   While the object was hovering above their car the driver stated that: "….Another thing I remember….as though I was light in weight and airy.   Something like the first time you experience an airplane takeoff or drop from an air pocket.   It felt like the air and everything was light and weightless."   She also reported an eerie silence or stillness while the UFO was near them.   Dr. Hynek investigated this case which is reported in Section I.   Her sense of weightlessness and the eerie quiet has been reported in a number of other close encounters.

Another interesting UFO close encounter occurred in Lemon Grove, California in November, 1973.   The witnesses were two eleven year old boys who accidentally encountered a landed UFO on a vacant lot in their neighborhood while out playing.   The details of this case can be found in Section I.   But the strange effect the object had on the boys as its "engine" revved up and it began rising vertically is described by the MUFON (Mutual UFO Network) investigator in this sentence: "They agreed that they felt as if they were going to black out and that they were running in slow motion."   This sense of being in slow motion has been reported in other close encounters around the world.

Even more bizarre are reports of light beams being bent in the vicinity of UFOs.   Section I includes two well documented cases of this type.   Reverend Gregory Miller of Norwood, Ohio along with many other witnesses on the night of October 23, 1949 observed a powerful searchlight beam aimed near a stationary UFO overhead actually bend about 27 degrees toward the object. Smaller UFOs apparently from the larger "mother ship" also caused the searchlight beam to bend when they were near it.   A similar case occurred near Burkes Flat, Victoria, Australia in 1966. In this case the headlight beams from an automobile were bent toward a brilliant UFO in a field adjacent to the road.

In addition to reports of weightlessness, objects being picked up and light beams being bent near UFO's there are also reports of clocks showing a loss of time.   Ted Phillips of MUFON investigated a case in Tennessee where the owner of a large collection of non-electric and electric clocks found all of them stopped after a UFO passed over his home.   Another well documented case occurred in 1979 involving Deputy Sheriff Val Johnson of Marshall County, Minnesota. Deputy Johnson experienced a close encounter with a brilliantly lit object that collided with his patrol car and caused both his car clock and wristwatch to both lose 14 minutes of time.

With UFOs causing weightlessness, a sense of being in slow motion, loss of time, lifting of objects and masses of water, and bending light beams toward the craft there seems to be considerable evidence of strong *attractive* gravitational fields around UFOs.   The puzzle is why UFOs would have an attractive gravity force, especially on their lower surfaces facing the earth? Shouldn't they be pushing away from the earth thus causing objects and people to experience a downward pressure when they happen to be under a low flying UFO?   If these machines use some kind of anti-gravity to levitate above the earth's surface then you would expect that masses below the craft would be pushed away, not pulled toward the UFO.   Yet the opposite is happening. This is baffling to say the least.   However as we will soon see this odd characteristic of UFOs is a critical clue to the technology used in these machines.

"Rippling" or "Fuzziness" and Partial Transparency of UFOs

A classic case involving the "rippling" effect on part of the structure of a UFO took place on March 29, 1952 over Misawa, Japan.   U.S. Air Force jets were performing practice intercepts when Lieutenant D.C. Brigham, the pilot of a T-6 trainer (used as the intercept aircraft), spotted an 8 inch diameter wobbling metallic disk closely following one of the F-84 interceptors.   Lt. Brigham reported that the edges of the disc, although presumably solid, appeared to "ripple". These smaller UFOs may be some kind of probes released by the larger UFOs for close observation of human activities.

Another "probe" encounter occurred in December, 1953 over North Korea during the Korean War. This sighting is detailed in the booklet "Advanced Aerial Devices Reported During the Korean War" by Richard Haines. A T-6 pilot, Lt. Barr, observed a somewhat triangular shaped craft which was about six feet long and 18 inches thick. When the object was about 600 to 800 feet from his aircraft, Lt. Barr noticed that the UFO appeared translucent, particularly on its outer edges. Sparks were also noticed a short distance behind the object.

A further example of this category occurred in the 1960's as described in Dr. J. Allen Hynek's book "The Hynek UFO Report". In this sighting the witness observed a bright oval shaped object somewhere between 40 and 50 yards in diameter. He described it as "having no distinct edges but rather a fuzzy outline….".

This fuzziness or rippling and partial transparency of what is thought to be a solid surface on UFOs is, I believe, another important clue to their technology.

Disappearing UFOs

The last characteristic of UFOs that we'll look at is a relatively rare behavior of these craft but nevertheless an important clue to how they function. On a number of occasions UFOs have been seen to disappear directly in front of witnesses. This is not to be confused with cases where UFOs are under observation but suddenly depart at such high speed that they are difficult for the eye to follow. The cases we are interested in here involve witness observations where the disappearance was unmistakable.

A very interesting case reported in Section I occurred on January 28, 1994 to Air France pilot J.C. Duboc and his crew who observed a red disc shaped UFO that as it began fading from view the witnesses were able to see right through parts of the UFO. Soon the entire UFO faded out. But moments later the UFO reappeared only to fade away and disappear a second time.

Another case occurred in 1931 over the Tasman Sea that separates New Zealand and Australia. The witness was pilot Francis Chichester who observed a UFO while flying in his Gypsy Moth aircraft. Chichester published a book titled "The Lonely Sea and The Sky" in which he described his encounter with the object: "It drew steadily closer, until perhaps a mile away when suddenly it vanished…. Then it reappeared close to where it had vanished…. It drew closer, and I could see the dim gleam of light on nose and back. It came on, but instead of increasing in size, it diminished as it approached! When quite near, it suddenly became its own ghost - one second I could see clear through it, and the next it had vanished." Where the airship had originally disappeared a diminutive cloud formed to the exact shape of the object and then dissolved.

A third well documented case involved Colonel Charles Halt and other witnesses at a U.S. Air Force Base in Great Britain in the winter of 1981. At one point in the series of events that comprised this sighting Colonel Halt described how a nearby brightly lit UFO disappeared before their eyes in a shower of sparks.

Disappearing UFOs are sometimes seen to shrink before they disappear as noted in the Francis Chichester sighting above. Another sighting in Columbia, Missouri demonstrates this effect as recounted in "The UFO Book – Encyclopedia of the Extraterrestrial" by Jerome Clark. On the night of June 28, 1973 a family observed an intensely bright 12 to 15 foot diameter object outside their mobile home. The trees near the object were being buffeted as if by a strong wind and the entire area was lit up like daytime. After a short time the trees stopped their frenzied motion and the witnesses were able to see that the object was of an aluminum color, brighter at the center and a dull white color toward the edges.   After moving around the area the object finally returned to its starting point about 50 feet away from the witnesses.   At this point the object started moving among the nearby trees and then to the complete surprise of the observers it began shrinking until it disappeared.

It is worth noting here that the fuzziness and partial transparency of UFOs discussed in the section before this one is very likely related to the UFO's ability to fade out and disappear completely.

Although there are only a limited number of cases where UFOs appear to "dematerialize" in front of witnesses, nevertheless this may be the most important clue of all in revealing the technology of these craft and the implications for modern physics.

Other UFO Characteristics

There are a number of other UFO characteristics that emphasize the strangeness of this phenomenon.   For example many witnesses talk about "solid light beams" emitted and then retracted "like a ladder' from UFOs.   Vehicle interference is another often reported effect in close encounters.   Examples of these effects can be found in Section I.

Solving the UFO Enigma
Chapter 3 in Section III

We are now ready to solve the UFO enigma.   As explained in the foreword of the book we are primarily interested in the UFO technology, i.e. how do these machines work?   The French government in its COMETA Report concluded that the best explanation for at least some UFO reports is that they are from advanced civilizations of extraterrestrial origin.   Other researchers have reached the same conclusion.   For our purposes here this will be our working assumption that these are real machines from civilizations beyond planet earth.

So what makes these UFOs tick?   The evidence points to strong attractive gravity fields around UFOs.   It would seem more logical that these craft are using some kind of anti-gravity. But if UFOs were using anti-gravity we would see objects and air near a UFO being pushed away. The opposite is true.   Repeated observations talk about water mounding under UFOs or dust and other objects swirling upwards toward the bottom surface of these machines.   This is puzzling. Anti-gravity would appear, at first glance, to be a sensible way to counter the earth's gravitational field.   But this does not seem to be the case.   Let's briefly review why anti-gravity is actually unlikely.

Quantum theory indicates that gravity is fundamentally different from the other three forces of nature.   The weak, strong and electromagnetic forces all have quantum spin of one and communicate multi-polar forces; i.e. weak hypercharge (W+, W-), color charge of the strong force (3 colors of red, green and blue) and positive and negative electric charge.   In contrast the postulated quantum particle for gravity, the graviton, has spin of two and the force associated with gravity is uni-polar and always attractive.   Gravity is an attractive force because matter is always gravitationally positive.   Compare this to matter particles that feel the electromagnetic force but can have different electric charges like the negative electron and positive proton.

What about negative matter?   (Not to be confused with anti-matter which is just normal matter with its charge reversed).   General relativity and Newtonian physics both theoretically permit anti-gravity in the presence of negative matter.   Unfortunately the Standard Model does not allow negative matter.   Nor has negative matter ever been detected in nature.   It looks unlikely that negative matter even exists.   Even if negative matter did exist its behavior would not lend itself to the kind of flight performances observed with UFOs.   Positive masses are repelled by negative masses.   But negative masses are attracted to positive masses.   The net result is a kind of Mexican standoff.   A UFO utilizing negative mass would need to continuously add or subtract negative mass from its structure in order to maneuver above earth's surface.   But there does not seem to be any credible UFO observations exhibiting this kind of technology.

Another problem with a pure anti-gravity propulsion system is what do you do to propel your craft when you are far away from a gravity field?   An anti-gravity system is fine while you are near the earth's surface but the deeper you go into space the less useful it is.

Although anti-gravity is a staple of science fiction it has little support from known science and even if it were possible it is only useful under certain conditions.   It is much more likely that UFO's use some technology that is applicable in a broad range of environments.   The bottom line is that anti-gravity is almost certainly not what UFOs use in their propulsion systems.

What about other possible propulsion methods such as countering the earth's magnetic field? Unfortunately the earth's magnetic field is so weak that levitating an object of even modest weight by creating a counter magnetic field would not be realistic.

This leads us back to the frequently observed strong attractive gravity field observed near UFO's.   I believe this is the decisive clue that is revealing UFO technology.   But if strong attractive gravity fields are at the heart of the UFO phenomenon then there are three essential questions which must be answered.   These are:

1) How does a UFO create a strong gravity field and what is the power source?

2) How is this strong attractive gravity field involved in the UFO propulsion system?

3) Does the strong gravity field have anything to do with how UFOs get to planet earth from other distant parts of our galaxy or universe?

We'll answer each of these questions in turn.

How UFO's Create Strong Gravity Fields

One of the key questions modern physics has been trying to answer is the hierarchy problem. The hierarchy problem actually has two aspects.   The first is the huge discrepancy in strength between gravity and the non-gravity forces.   Put another way, why is gravity so weak?   The second aspect of the hierarchy problem is the vast difference in mass between the Higgs boson associated with the weak scale and the Higgs boson associated with GUT theories even though both Higgs are in the same family.   For our purposes here we are interested only in the weakness of gravity aspect of the hierarchy problem.

The weakness of gravity looks like it can be explained by recent theories that predict the existence of extra dimensions beyond the four of traditional space-time.   The basic idea of these theories is that gravity is weak because the vast majority of gravitons that carry the gravitational force are in the extra dimensions.   This is why gravity appears so weak in our 4-D world.   In string or membrane theories gravitons are considered to be closed loops which means they can travel anywhere in the extra dimensions.   On the other hand many theories regard non-gravity matter and force particles as open strings with both ends attached to the membrane we live on. These strings, according to these theories, cannot leave our membrane to wander into the extra dimensions.

As previously mentioned one of the most interesting of the extra dimension membrane theories is Randall-Sundrum 1 or RS1.   As we saw RS1 postulates a universe with two branes separated by a 5th dimension called the bulk.   Although RS1 predicts or needs only a 5-D universe it could have additional curled up dimensions too.   [In this way it might be associated with 11-D E(8)xE(8) theory.]   One of the two branes is called the TeV brane reflecting the low energy or masses of particles on this brane.   This is where we live.   The other brane is called the Planck brane where energies/masses are believed to be in the Planck scale range.   RS1 differs from other extra dimensional theories in that it theorizes that the extra dimensional bulk is severely warped, primarily by the energy of the Planck brane.   This warping causes masses and energies of particles to increase exponentially as they leave the TeV brane and approach the Planck brane.   It also causes most gravitons to be concentrated or localized in an area near the Planck brane.   The TeV brane has energies/masses in the 10^3 Gev range (equal to one TeV) while the energy/mass scale near the Planck brane is around 10^{19} GeV or a difference of around 10^{16} GeV!   This is huge.

Whenever you have extra dimensions you have the possibility of KK particles.   The interesting thing about RS1 is that the KK particles of the 4-D graviton can have enormous interaction strengths.   How strong?   The 4-D graviton KK partners from the bulk, should we produce them in the Large Hadron Collider (LHC), will have strengths 16 orders of magnitude greater than ordinary 4-D gravitons!

The LHC will smash hadrons together in head on collisions.   Hadrons are particles made up of quarks such as are found in protons or mesons.   Beams of quarks and anti-quarks will be accelerated to sizeable fractions of the speed of light and then collided head on with each other. KK particles will be produced from these collisions assuming the extra dimension theories are correct.

KK particles result from quantum mechanics equating particles with waves.   These waves can only oscillate an integer number of times in an extra dimension.   Otherwise the waves will overlap and self destruct.   So if the extra dimension is a curled up circle the wave can do one full oscillation, two full oscillations, three full oscillations, etc.   More oscillations is equivalent to higher frequency.   We know from quantum mechanics that $\mathbf{E=hf}$ so the higher the frequency "$\mathbf{f}$" the greater the energy.   The higher energy KK gravitons will be seen in our world to have mass. The greater momentum of these KK particles translates into mass.   However the lightest KK gravitons will have no momentum from the extra dimension and will therefore be massless.

In RS1 the natural scale of the TeV brane is around one TeV.   Particles in the bulk close to one TeV such as 1 TeV, 2 TeV, 3 TeV, etc. will tend to be numerous near the TeV brane.   The LHC should produce particles in this range including KK gravitons.   These KK gravitons will interact gravitationally some 16 orders of magnitude greater than ordinary gravitons. Mathematically this is 10^{16} times stronger than the gravity we are familiar with.   They will be very noticeable!   RS1 calculates that these KK gravitons will decay inside the LHC into other detectable particles from which scientists can deduce that the original particle was a spin-2 KK graviton.

Here is the probable source of the powerful attractive gravity fields observed around UFO's. Other civilizations beyond earth have likely figured out how to create sustained and enormously powerful gravitational fields using KK gravitons.   This would then explain why UFOs always seem to have a circular aspect in some part of their configuration.   These machines are particle accelerators.

But if we earthlings need to create an accelerator 17 miles in circumference like the LHC in order to create KK gravitons then how could other civilizations manage to do it in a craft often only 100 feet in diameter or less?   I believe this is an engineering problem that can be figured out.

Let's look at some of the engineering challenges the UFO builders must have encountered. One of the key engineering problems is synchrotron radiation produced by circular particle accelerators.   Particles moving in a curved path experience angular acceleration and lose energy in the form of synchrotron radiation.   To compensate for this accelerators must supply additional energy during operation.   Building the machines with a larger diameter also helps decrease the acuteness of the track's curvature and reduces the energy loss as was done with the LHC. However in an accelerator producing an immensely strong gravity field the parameters change. The donut shaped strong gravity field created by the UFO curves space around the rim area of these machines.   As we learned in general relativity objects or even light rays always follow the curvature or geodesic of space-time.   So as far as a subatomic particle moving around the rim area of a UFO is concerned it does not see this space as circular and therefore does not experience angular acceleration.   This is apparently one reason why UFOs can be relatively small because the physics begins to change in strong gravity fields.   I am sure we will be able to figure out the various engineering problems as we begin to create strong gravity fields ourselves.

It is also important to ask what will be the characteristics of the gravity field created by particle accelerators.   We know from quantum theory that bosons with mass interact only over short distances.   This is known to be true of the non-gravity bosons.   But we know that RS1 and other extra dimensional theories predict that both massive and massless KK gravitons will be produced by the LHC.   Therefore any strong gravity field created should have both long range and short range characteristics.   However there may be other factors that we need to consider too which we'll discuss shortly.

In the LHC there will be four locations around the particle accelerator rim where particle collisions will take place.   The UFOs would have many more locations where these collisions take place to produce a more comprehensive gravity field.   The particle collisions may take place sequentially which could explain the often reported "wobble" of UFOs at "low" power when they hover in a stationary position.   We'll see how UFOs "levitate" shortly.

Another question is what is the power source that is used to operate the particle accelerator? My guess is the UFO builders have figured out how to create controlled fusion using strong gravity fields.   Again I am sure that we will solve the engineering problems associated with compressing perhaps hydrogen gas to produce fusion energy utilizing strong gravity fields.   In fact 'artificial' gravitational compression may be easier than using our current method of electromagnetic confinement of an ionized gas in creating fusion power.

There is some circumstantial evidence that UFOs house particle accelerators.   Historically there have been reports of a gossamer like substance called 'angel hair' that has been seen falling to earth after UFOs have passed overhead.   An angel hair sample was collected in Sonora, California in 1976 and analyzed for its content.   Among the substances detected in the sample was a radioactive isotope of hydrogen called tritium.   Tritium is rare in nature.   It is only produced in significant amounts by humans in nuclear reactors or particle accelerators.   More samples of angel hair would be required to confirm this finding but it is nevertheless intriguing.   This particular analysis of angel hair was reported in the Fall 2001 International UFO Reporter, Volume 26, Number 3.

But now we are left with the puzzle of how strong attractive gravity fields created by UFOs could possibly be useful in counteracting earth's attractive gravity?   Using attractive gravity to counter the earth's attractive gravity does not seem to make sense.   We'll tackle this question next.

How Strong Gravity Fields Are Used in UFO Propulsion

For a UFO to perform the often phenomenal aerobatic maneuvers that have been observed many times in earth's atmosphere it requires at least two essential factors.   These are:

1) The UFO has to minimize or eliminate the effect of earth's gravity on its mass.

2) The UFO must minimize or eliminate the effect of inertia on its structure.

There are other factors that are important for UFOs too such as avoiding air friction during high speed flight but these considerations are secondary to the two fundamental factors above.   Once we deal with the two essential issues above we'll review other factors too.

So how does a UFO minimize or eliminate the effects of gravity and inertia?   We'll tackle the problem of UFOs dealing with earth's gravity first.

1) How UFO's Minimize the Effect of Earth's Gravity

We have evidence that UFOs have considerable mass.   The case of the UFO landing on railroad tracks in France leaving indentations in the wooden ties was analyzed and the craft's weight was estimated to be thirty tons.   Considering that it was only the size of a Cadillac suggests that UFOs are very massive objects.   Other studies such as one done by Ted Phillips, a UFO researcher, also indicate that UFOs are machines of great mass.   Probably this results from these craft being essentially particle accelerators powered by fusion reactors.   Yet somehow these massive objects are able to levitate above our planet's surface as if gravity has little effect on them. How do they do it?   The answer I suspect is surprisingly simple.

I believe that the powerful attractive gravity fields created by UFOs eliminate most of the effect of earth's gravity on these craft.   They do this by creating a kind of graviton "sink" envelope around the UFO or at least on its bottom and side surfaces during "low power" hovering.   The graviton sink results from the UFO accelerator producing powerful KK 5-D gravitons which in turn create micro-black holes.   These micro-black holes are predicted to occur in the LHC if RS1 theory is correct.   The most intense part of this graviton sink envelope is probably only a few angstroms thick and located just outside the surface of the machine.   Virtual gravitons emitted by the earth or by the UFO are simply absorbed by the micro-black holes.   When extra dimensions are involved M-theory based calculations indicate that no gravitons are emitted onto the TeV brane during the decay of the micro-black holes. 5-D gravitons are thought to be emitted but only into the 5-D bulk.   This means that little or no exchange of gravitons occurs between the UFO and planet earth.   The force of gravity is for the most part not communicated to the UFO.   The UFOs probably retain a small amount of "weight" in order to allow for action-reaction thrusters used for maneuvering in our atmosphere.

The numerous symmetric portholes often seen around the rim area of UFOs may be where subatomic particles are ejected from the accelerator to collide with one another just outside the craft's surface. This would explain why UFOs are frequently seen to be surrounded by a brilliant glow.   This results from air, dust and other debris being sucked into the graviton sink at very high speed thus imparting energy to these air and dust molecules which causes them to emit a spectrum of radiation.   Micro-black hole decay also contributes to this radiation.

But when UFOs are moving slowly you can often clearly see their metallic surfaces.   At this 'low power' setting the UFO may have its particle accelerator collisions taking place at an alternate location just inside the skin of the outer rim area.   The graviton sink may not be as intense but it would be strong enough to keep the craft levitated.   The low power setting allows the UFO operators to easily observe their surrounding environment without the interference of the high power setting radiation glow outside the machine.   However there are certain circumstances, as we shall see, where having the graviton sink just outside the surface of the craft is essential.

The "light" weight of these objects would explain why the outer rims are often seen rotating. This provides stability to the craft via centrifugal force in exactly the same way as does spinning a light weight Frisbee when we toss it across the yard. Since the UFO is so light as far as the earth is concerned it would only take low mass particles to propel the machine in different directions. Particle thrusters could be located symmetrically on the rim and on the bottom surface. Only modest amounts of thrust would be necessary so these particle thrusters would not be too noticeable. The thrusters would need to be located above the area of the graviton sink or else the sink would swallow the particles from the thrusters. Alternatively the gravity field might be adjusted at the thruster location to allow for the expulsion of low mass particles. Since some KK gravitons will have mass that means the gravity field they produce will be short range and could possibly be manipulated.

Once the UFO is a reasonable distance above the ground it would no longer swirl tree tops or pull dust, water or other material toward or into the graviton sink. At night and sometimes in daylight UFOs are often seen to glow. As noted above this may be from air molecules accelerating into the graviton sink (made up of micro-black holes) and producing electromagnetic radiation including highly energetic gamma rays. Also when black holes evaporate they are thought to emit bursts of gamma rays. Of interest in this regard is that in Section I, Chapter 2 'The Classic UFO Cases' doctors concluded that prospector Steven Michalaq was briefly exposed to a blast of X-rays or gamma rays in his close encounter with a landed UFO in Classic Case number 10.

But the gravity field would drop off in strength fairly quickly. This artificial gravity field/sink need only be strong in a very thin envelope to absorb virtual gravitons into the micro-black holes. The fact that extra dimension theories are theorized to produce strong gravity and micro-black holes at just a few TeV in the LHC means the possibility of creating a graviton sink is not so far fetched. The major problem of minimizing the earth's gravity is accomplished by the graviton sink and relative to earth's gravity field the UFO moves about with ease.

Another issue is whether we can really create a graviton sink that will absorb gravitons? The graviton sink around the UFO is essentially a collection of micro black holes. In a strictly 4-D world producing micro-black holes in earth based accelerators would be impossible. For a 4-D world it has been estimated that to produce the minimum energy of 10^{19} GeV within the Planck length of 10^{-33} centimeters, which is needed to create a black hole, that it would require an accelerator 1,000 light years in diameter. A little beyond our current technology! But in many extra dimension theories (including RS1) it is believed that the Planck mass where gravity becomes strong like the other forces can be reached at energy levels merely in the TeV range. The LHC is designed to go as high as 14 TeV with proton-proton collisions. Thus creating the micro-black holes for the graviton sink looks entirely plausible.

But what happens to the gravitational energy of say a forty ton UFO hovering over the earth's surface? This is, after all, a considerable mass to keep suspended above the ground. The continuously created micro-black holes do absorb the gravitons from both the earth and the UFO but then these micro-black holes quickly decay too. If most of the decay energy from these micro-black holes that is needed to suspend the forty ton craft is emitted into the earth's atmosphere then UFOs would be sources of very large amounts of decay radiation. However this does not usually seem to be the case.

An interesting paper published in 2006 titled "Black Hole Particle Emission in Higher Dimensional Spacetime" by several American physicists and an Italian physicist may explain what happens to most of the gravitational energy and other forms of mass and energy absorbed by a rotating micro-black hole. In their paper the authors explain quantitatively the types of particle and

EM radiation emissions from decaying black holes and where it ends up. Their paper is based on theories that utilize extra dimensions such as Warped Geometry RS-1. It turns out that in theories with eleven space-time dimensions [for example 11-D E(8)xE(8)] almost 99% of the decay emission energy from a micro-black hole ends up in the 5-D bulk in the form of 5-D gravitons. In effect the gravitational energy (and other forms of mass/energy absorbed by the micro-black holes) of the forty ton UFO in our example is mostly shunted into the 5-D bulk. The remaining one percent of energy is emitted onto the 4-D TeV brane where we live in the form of scalars (Higgs particles), fermions and gauge bosons. Included in the gauge bosons would be gamma rays that have often been reported near UFOs. The decay onto the TeV brane should resemble something called Hawking radiation.

We will use known particles like protons in the LHC to create KK gravitons and most likely micro-black holes too. But advanced civilizations may be using other types of particles such as the supersymmetric partners if these entities exist. If supersymmetry is correct the LHC should give evidence of this too.

The graviton sink field may not need to be that strong under ordinary conditions in order to get the job done. The reason for this is that the number of virtual gravitons emitted by earth per unit of area is miniscule in comparison to say the virtual photons exchanged between charged particles per unit of area. Gravity is extremely weak versus the other forces. In the extra dimension theories most gravitons are thought to be in the bulk.

But there may be situations where making the gravity field/sink much stronger and with the particle collisions taking place in an envelope outside the UFO's surface is useful. For example during the Korean War a brightly lit UFO was seen to approach a battlefield and head directly toward exploding aerial artillery shells as if it were some kind of game. Under these circumstances the UFO probably cranks up the strength of the outside gravity field/sink and the shrapnel from exploding shells is either deflected by the curved space near the UFO or effectively shunted into the bulk. The brilliant illumination surrounding the UFO is an indication that the outside graviton sink is in operation. The details of this sighting can be found in Richard F. Haines booklet titled "Advanced Aerial Devices Reported During the Korean War".

In another incident bullets from a fighter plane were seen to follow a curved path around the UFO then straighten out beyond the object. This too could be an example of material objects following the strongly curved space-time geodesic around the UFO.

At other times witnesses have fired bullets at UFOs and distinctly heard them impact the metallic surface. This probably occurs only when the UFO does not have the gravity field/sink cranked up to higher strengths in the area outside the surface of the machine but instead uses a weaker, shorter range graviton sink located inside the craft. The weaker inside graviton sink is needed when the craft is near or on the ground to avoid too much disturbance of the environment.

In some UFO close encounters witnesses report a powerful suction sound or sensation. This would result from the graviton sink sucking in air and dust molecules as previously mentioned. Witnesses also report intense gravitational effects such as clocks stopping, a sense of weightlessness or the feeling they are in slow motion. This too would be a side effect of the strong gravity field created by the KK gravitons.

An extraordinary UFO sighting occurred in the 1970's where the witnesses apparently observed both the outside and inside graviton sink in action. The details of this sighting can be found in the 2002-2003 winter issue of the International UFO Reporter in an article by UFO researcher Michael D. Swords. Two young men in the early 1970's working at a resort in Death Valley, California encountered a glowing red ball as they walked back to their rental house in the wee hours of the morning.   As the ball moved toward the base of a nearby mountain it seemed to create a vortex underneath which caused hundreds of rocks, some as large as melons, to rise up off the ground in a swirling column directly under the luminous ball.   Even as the object gained altitude it pulled along this rotating column of heavy rocks beneath it.   The only sound was that of the rocks clicking against each other.   But then the light "blinked out" and the huge column of rocks came crashing down the mountainside spilling over onto the road.   The object then "blinked" its luminosity back on again and moved away into the distance minus the rocks.

The young men very likely observed both the powerful outside graviton sink when the ball was lit up and the weaker inside graviton sink when the object's luminosity was turned off.   Clearly the spherical UFO was trying to rid itself of its uninvited guests.   The reason the column of rocks was rotating under the craft is probably due to the particle collisions in the accelerator taking place sequentially thus causing the UFOs strong gravity field to create a twisting action.

One additional UFO sighting clearly demonstrates the powerful *attractive* gravity field surrounding UFOs.   In the November 4, 2010 National UFO Examiner edited by Roger Marsh there is a remarkable nighttime sighting of two triangular UFOs in California that occurred on October 25, 2009.   The witness reported that the "sky and stars in that area (i.e. around the UFO) were distorted kind of like they were being pulled towards the center of the UFO….".   This gives a good indication of the enormous strength of the attractive gravity field around these machines, literally bending the light beams from stars behind it..

Throughout this discussion on how UFOs minimize the effect of earth's gravity I've used the term 'graviton' to describe the quantum entities emitted by earth (and the UFO) that carry the gravitational force.   However it is actually virtual gravitons that carry the gravitational force just as virtual photons carry the electromagnetic force.   I used the term 'graviton' just for brevity.

So a strong gravity field/graviton sink can be used to explain how UFOs levitate above earth's surface as well as some other interesting behavior of these craft.   This assumes that gravitons predicted by quantum theory are real.   Recent theories like M-theory and its derivatives also support the existence of gravitons.   As far as the pre-quantum general relativity description of gravity is concerned we can say that the graviton sink envelope around the UFO modifies the earth generated warped space-time in the area occupied by the craft.   But eliminating the effect of earth's gravity does not necessarily explain how UFOs appear to defy inertial forces as evidenced by their rapid accelerations and sudden changes in direction.   To explain how they overcome inertia we have to do a little more work.

2) How UFO's Minimize the Effect of Inertia

What is inertia and where does it come from?   Inertia is defined as the resistance of a body to acceleration.   But the origin of inertia is a mystery.   Mach believed that the inertia of an object was due to the gravitational attraction of all the other matter in the universe.   Einstein referred to this as 'Mach's principle'. However this would require a feedback mechanism that would work instantaneously but this contradicts special relativity which forbids any signal going faster than the speed of light.

Einstein had hoped to include Mach's principle in his general theory of relativity but eventually concluded it could not be done.   Nevertheless Einstein did believe that inertia in some way had its origin in gravity.   This is reflected in his equivalence principle that equates gravitational and inertial masses.   Note that the equivalence principle is a postulate, i.e. it is a supposition, but it is backed up by numerous experiments.   The equivalence principle does not yet have a theoretical foundation.

In order to understand how UFOs minimize inertia we must first determine the origin of this force.   Sometime before theories involving extra dimensions became acceptable in the physics community I had toyed with an idea that might explain inertia.   We'll take look at this next.

The Oscillation Theory

My original idea to explain inertia was what I called the *two spacetime theory*.   This was before M-theory and its concept of membranes.   The basic idea of the two spacetime theory is that all quantum particles oscillate between a foreground spacetime (FST) which is the world we live in and a background spacetime (BST) that is not directly visible to us.   The BST is connected to all points in the FST.   I speculated that deep inside the BST time was stopped or possibly went in both directions.   This meant that particles oscillating into the BST are instantaneously connected to all parts of the FST during each oscillation.   In addition I assumed the BST had a great deal of energy. This I thought could be the origin of inertia and possibly what we call mass.

The Higgs field may actually be the BST.   As quantum particles oscillate into the BST they pick up what we call mass or energy.   This is the origin of a particle's 'Higgs' mass/energy.

A particle moving at constant speed oscillates between the two spacetimes in a uniform manner and maintains the same mass/energy content.   However if a particle is accelerated in the FST it has to cover a greater distance between oscillations in both the FST and BST.   More distance in the BST exposes the particle to more of the energy content of the BST and the particle gains mass/energy.   We see this in the FST as requiring more force to accelerate an object.   We call this inertia.

But the idea of a background spacetime was pure speculation.   However with the development of M-theory and the hierarchy solving theories that it spawned, along came just the theory I needed for my two spacetime idea.   This theory was of course Randall-Sundrum 1 (RS1).   The reader will notice that I already used RS1 to explain the strong gravity fields around UFOs.   Is it a coincidence that RS1 also looks like it is involved in explaining inertia?   Probably not.   It appears that all of this is tied together.

In RS1 the bulk with its tremendous gravitational energy is the BST of my two spacetime theory.   My original theory could not explain the energy of the BST or why time would slow down or nearly stop.   But with the exponential increase in the number of bulk gravitons as one approaches the gravitybrane in RS1 both the enormous energy and slowing of time are both explained.   RS1 is exactly what I needed.   I've since renamed the two spacetime theory to simply the *oscillation theory*.

If all particles in the universe oscillate between the TeV brane and the bulk then this could explain the equivalence principle of general relativity.   Inertial mass and gravitational mass both arise from the same strong gravity field in the bulk which is why they are closely related.

One important question is why do particles oscillate between the TeV and bulk branes?   The answer may be that all quantum particles are essentially individual mini-black holes.   Physicists have noted that black holes and elementary particles are very similar.   Both are described by their mass, spin and charge.   Some physicists have even speculated that elementary particles are the equivalent of tiny black holes.   If this is so then as a particle oscillates to the TeV brane its energy may be too large for it to stay there so it quickly pops back into the bulk.   As it travels deeper into the bulk its energy becomes again so large that it pops back to the TeV brane.   And so on.

Different types of strings pick up different levels of energy in the bulk based on how deeply they penetrate the bulk.

A 2009 paper by two researchers, Donald Coyne from UC Santa Cruz (now deceased) and DC Cheng from the Alamaden Research Center, proposes the possibility that subatomic particles are varying forms of stabilized black holes. Interested readers can see a write-up of this idea in the MIT Technology Review of May 14, 2009.

So how do we use the particle oscillation idea in determining how UFOs overcome inertia? Here's where it gets interesting. The same strong gravity field created by the UFO is applied to most of the mass of the craft. This strong gravity field slows down or even halts the oscillation of the quantum particles making up the UFO. As the oscillation is slowed down or stopped the craft, for the most part, no longer feels the inertia which comes from the bulk. This may be how UFOs eliminate inertial effects on their craft.

Of course if the bulk is also responsible for the 'Higgs' and thus the mass/energy of quantum particles then these particles making up the UFO would no longer have their 'Higgs' induced mass. This would be a serious problem since if all quantum particles in the UFO lacked Higgs mass then various quantum processes would be altered and the UFO would disintegrate!

We need to digress here a little to explain why a UFO would disintegrate without its Higgs (or oscillation) induced mass. Higgs gives mass to the W+, W- and Z neutral bosons but not to the photon. These four bosons in their massless state are involved in the electroweak local gauge symmetry. Once the W+, W- and Z neutral bosons gain mass the electroweak symmetry is broken into the weak force and electromagnetic force. The weak force is carried by the W+, W- and Z neutral bosons which now all have mass. Since fermions (matter particles) feel the weak force this results in quarks and leptons (including electrons) gaining mass. However most of the mass of a proton (or neutron) is due to relativistic effects from the motion of the quarks confined inside the proton as well as from the color gluon fields that bind the quarks together. Something like 98% of a protons or neutrons mass comes from these relativistic effects. Only around 2% of the nucleon's mass comes from the Higgs field.

So even if the Higgs (or oscillation) disappeared atoms would still have most of their mass from the relativistic effects produced by the strong force. In fact without the Higgs field the strong force would even induce a small Higgs-like mass in the W+, W- and Z neutral bosons while leaving the photon massless. Quarks and leptons would via the weak force (which they feel) also gain a small Higgs-like mass. But therein lies the problem! The smaller mass of the W+, W- and Z neutral bosons and the reduced mass of the electron will upset the apple cart!

Radioactive beta decay of nucleons results from the weak force. But if the weak force bosons had much lower masses then the weak force would operate over a longer range within the nucleus and therefore much more frequently. Beta decay would happen millions of times faster. In other words the nucleus of atoms would disintegrate quickly along with the UFO and its (gulp) occupants!

Lower mass electrons also cause a problem. The radius of an atom is inversely proportional to the mass of the electron. With lower mass the radius of the atom increases significantly which ultimately leads to unstable molecules.

No matter how you slice it without the correct Higgs (or oscillation) portion of an atom's mass our UFO is in serious trouble! Fortunately there may be a way around this problem. If quantum particles get their mass and inertia from oscillating into the strong gravity field of the bulk then maybe the strong gravity field created by the UFO itself can be adjusted to provide just the correct Higgs induced mass that is needed. But note that the inertia resulting from this technique would be local, i.e. just from the UFO itself.   The inertia from the entire universe would no longer pull on the UFOs structure.

Now we can easily explain the rapid accelerations and hairpin turns that UFOs are seen to perform.   Because of the graviton sink around the UFO the warped spacetime caused by earth that is in the vicinity of the UFO no longer "sees" the craft.   Then the UFOs elimination of external inertia using a strong gravity field to slow or stop oscillations of its constituent quantum particles allows the craft to accelerate with essentially no inertial effect.   Using the same artificially created strong gravity mechanism the UFO has overcome both the earth's gravity and the universe's inertia.

As a bonus the strong attractive gravity field around the UFO pulls layers of air molecules in a kind of cushion around the UFO.   The further away from the UFOs surface the air molecules are the less tightly they are held.   This air cushion gradient allows the UFO to travel at supersonic speed without creating a shockwave or causing the craft's surface to heat up from friction.   The skin of the UFO is not directly hitting the air molecules as the craft speeds through the atmosphere. Instead each successive layer of air molecules is going a little slower as you get closer to the UFOs surface.   Finally at or near the UFOs surface there is little or no movement of air molecules (or dust or water drops etc.).

Since molecules from the atmosphere are either barely moving or not moving at the UFO's surface then there is no friction caused heating of the craft's skin.   No compression wave is formed either because the kinetic energy is dissipated over successive layers of air molecules in the UFOs air cushion.   Therefore there is no sonic boom.   One of the key pieces of evidence that UFOs are surrounded by a strong gravity field is the lack of compression waves or friction caused heat. Other UFO researchers have commented on this too.

One other point needs to be mentioned here.   What is the difference between a subatomic particle that is theorized to be a mini-black hole and a micro black hole created by the LHC (and UFOs)?   The micro-black hole of the LHC (or UFO) is far more energetic than the ordinary subatomic particle that is thought to be a stabilized black hole.   The LHC will smash subatomic particles together at near light speeds in head on collisions which will produce micro-black holes of tremendous energy.   It is the enormous energy of these LHC micro-black holes that should give them the ability to shunt virtual gravitons into the bulk.   Of course the LHC type black holes are unstable and will disintegrate quickly so UFOs must continually generate new micro-black holes in order to sustain their graviton 'sink'.

Now we see how UFOs use strong gravity fields as the central feature of their propulsion systems.   It makes sense that UFOs would optimize the use of a single technology for the four separate functions of minimizing inertia, largely eliminating the effect of earth's gravity field, creating fusion power and providing a protective cushion of air around the craft.   Are there other functions where strong gravity fields might be useful too?   Yes.   There is one more essential function that the strong gravity field likely performs.   This is the question of how UFOs get to planet earth in the first place.   We'll solve this last piece of the UFO enigma next!

How Do UFOs Get to Planet Earth?

The reader may have noticed that if the oscillation theory is correct then at any given moment only half the constituent particles that we are made of are actually on the TeV brane where we live. So is there another mirror image of ourselves in the bulk at any given instant?   No because time has come to a standstill deep inside the bulk so the quantum particles that make us up are everywhere in the bulk and in contact with all other particles in the universe. This is where inertia and mass come from. Without time there is no speed. So it takes no time for a particle to go from point A to point B or point B to point C deep in the bulk. In other words all particles deep in the bulk are everywhere at once. I'm assuming that individual quantum particles oscillate deep into the bulk where RS1 has time either at or near a standstill.

How big is the bulk? A type of string theory called T-duality says the physics of a universe compactified in a tiny microscopic universe can be identical to the physics of a large universe such as the one we live in. Could it be that the oscillation of quantum particles, if correct, is between T-dual universes? Our large universe may be T-dual to a tiny universe which is the bulk/gravitybrane of RS1. The Heterotic B E(8)xE(8) string theory in 11 dimensions may represent this T-dual (double) universe which in turn could be the basis of RS1.

So how might a UFO take advantage of the bulk to travel between stars that are huge distances apart on the TeV brane? I believe that they use their strong gravity field to slow down and synchronize the oscillations of all the quantum particles in their craft including those of the occupants. By adjusting the oscillations carefully they can move all their quantum particles over to the bulk simultaneously. They would then disappear from the TeV brane. This would explain the interesting cases where UFOs are seen to fade out, become partially transparent and then disappear right in front of witnesses. In some cases the UFOs are seen to shrink as they disappear. If RS1 is correct then objects moving into the bulk would be expected to shrink at least somewhat.

But once in the bulk wouldn't the craft disintegrate as its quantum particles are dispersed everywhere in the bulk? No. This is where the strong gravity created by the UFO shields it from being torn apart and dispersed in the bulk. The strength of the gravity field needed to maintain the integrity of the craft is a function of how far the UFO goes into the bulk.

How far does the UFO go into the bulk? In RS1 the number of gravitons (and hence the force of gravity) increases exponentially the further you move away from the TeV brane. The craft might only have to go far enough into the bulk for time to slow down significantly but not completely stop. At this point the distance in the bulk that corresponds to the distance between stars in the TeV brane would be very small. In RS1 the further you go into the bulk time goes slower and objects (and distances) get smaller. The UFO would be protected by its strong gravity field (adjusted to be stronger than the bulk gravity at that point) and the integrity of the craft would be protected. The craft could now travel the short bulk distance between stars that are on the TeV brane.

However while the distance factor has shrunk the time factor has dilated or expanded which means that the time needed to travel even a short distance in the bulk would be enormous. But since the UFO operates by "swallowing" gravitons into its graviton sink it could eliminate gravitons in the direction it is traveling and in this way eliminate the expanded time factor too. A UFO coming from Alpha Centauri, which is about 4.3 light years from earth, might only take half an hour in the bulk to reach our planet. Upon reaching the position in the bulk that corresponds to planet earth the UFO would adjust its oscillations to pop back onto the TeV brane.

Navigating the bulk is a skill each advanced civilization would learn over time. Perhaps when the UFO is only part way into the bulk it can detect the gravity of stars on the TeV brane at different locations in the bulk. Eventually a 3-D map of all star locations in the bulk corresponding to positions on the TeV brane would be developed.

The key to fast interstellar travel is gaining access to the bulk. Craft from other civilizations that we call UFOs appear to have done just that. With this proposed explanation of how UFOs get to planet earth we have completed the last piece of the UFO enigma at least as far as their technology is concerned.

In the conclusion for this discussion of UFO technology we'll also see how the proposed physics used to explain these alien craft may offer new insights into our understanding of physics here on earth.

<u>Conclusion on UFO Technology</u>

In solving one of the biggest puzzles in modern physics, the gravity related hierarchy problem, we have found a likely solution to the UFO enigma.   M-theory and RS1 with their branes and warped geometry look to be the keys to understanding the physics of the UFO phenomenon. Then by making just one basic modification to known physics, i.e. the oscillation theory to explain inertia and particle mass/energy, we may have the last piece of the puzzle needed to replicate UFO technology.

The UFO evidence points strongly toward the existence of powerful attractive gravity fields around UFOs.   For example a driver in Australia reported that his headlight beams bent *toward* a nearby UFO.   Space-time near a UFO would have to be significantly warped for this to occur. This case and others tell us that UFOs are using intensely strong attractive gravity fields.   The immense strength of these fields is what makes the micro-black holes graviton sink surrounding the craft a plausible idea.   Using the same strong gravity field to dampen or eliminate inertia, assuming the oscillation theory is correct, makes perfect sense.   Synchronizing the oscillation of all the UFOs constituent particles by again using the strong gravity field gives the craft access to the bulk where the warped gravity physics allows for the equivalent of fast interstellar travel on the adjacent TeV brane.

The bottom line is that strong gravity fields are almost certainly the basis of UFO technology. If upcoming experiments at the Large Hadron Collider give us evidence to support RS1 theory (or something similar) then I believe we will be taking a critical step toward reproducing UFO technology here on earth.   Finding evidence to support the oscillation theory will likely require further research into strong gravity fields.   Once physicists produce evidence of strong gravity you can be sure that numerous experiments will follow.

It may be possible to test the hypotheses for this proposed UFO technology by gathering additional data from UFO sightings and encounters.   For example specialized equipment could be used to record spectroscopic data during UFO sightings.   Micro-black holes occurring in the atmosphere (i.e. around the UFO) should have a characteristic signature that would be revealed in the spectrographic data.   Bursts of gamma rays would be one possible indicator of micro-black holes around a UFO.   Hawking radiation (named after English physicist Steven Hawking) might be produced as the micro-black holes rapidly evaporate.   Hawking radiation would produce a black body spectrum of radiation due to quantum mechanical effects.

Beyond solving the UFO enigma; M-theory, RS1 and the oscillation theory if they are confirmed may resolve some other major puzzles in physics and explain the basis of many physics 'principles'.   I should mention that M-theory is not a complete theory yet but perhaps LHC experiments will help to resolve remaining issues while at the same time verifying its basic concepts.

In the next section we'll take a look at how our solution to the UFO enigma may give us new insights into modern physics.

The UFO Solution Applied to Physics on Earth

If branes, the bulk and warped gravity really exist (perhaps based on 11-D E(8)xE(8) M-theory) then along with the oscillation idea we may have a way to explain a number of foundational principles and unsolved questions in modern physics.   We already saw that inertia could be explained by the existence of the bulk's warped gravity/geometry and the oscillation theory.   But what about other areas of physics such as non-locality (or entanglement), quantum tunneling, the equivalence principle, special relativity, speed of light, masses of the fundamental particles, Higgs, etc?   Can we gain a better understanding of these fundamental concepts of modern physics from our solution to the UFO enigma?   I think we can.   Let's see how we may be able to shed new light on our current understanding of physics.

Non-locality

Non-locality refers to how quantum particles can be affected instantaneously by events that occur far away thus apparently violating special relativity.   The slowing or halting of time in the bulk along with the oscillation theory may explain these instantaneous quantum effects.   Quantum particles during their oscillation cycle into the bulk can 'touch' distant parts of the universe instantaneously.   Special relativity would not be violated on the TeV brane because the instantaneous effect would take place in the bulk but is then communicated back to the TeV brane. The exact details of how two particular particles are entangled needs further explanation but the potential mechanism for achieving this is RS1 and the oscillation theory.

Quantum Tunneling

The quantum effect known as tunneling may also be explained by RS1 and the oscillation theory.   Tunneling says there is a certain probability that a quantum particle will appear on the other side of a supposedly impassable barrier.   For example in radioactivity a proton with high momentum (energy) will occasionally "jump" outside the range of the strong force and leave the atomic nucleus.   In the oscillation theory the energy (momentum) of a particle is a function of how far it oscillates into the bulk.   A particle with more momentum oscillating deeper into the bulk may be more likely to be affected by predictable disturbances in the bulks gravity field.   The bulk's gravity field is impacted more or less instantaneously deeper in the bulk.   A cosmic event such as a supernova can cause a disturbance in the bulk.   Since there is massive energy deep in the bulk such disturbances will have only a modest effect.   However the modest disturbance of the bulk may be sufficient to deflect a high energy proton oscillating into the bulk just enough that when it emerges back onto the TeV brane it is outside the atomic nucleus.   This may be how quantum particles "jump" seemingly impassable barriers in a statistically predictable fashion.

Special Relativity

The Lorentz-Fitzgerald contraction, which Einstein incorporated into the special theory of relativity, may be explained as a result of the oscillation of quantum particles between the TeV brane and the bulk.   Length contraction in special relativity is a consequence of two things: 1) the speed of light being constant and 2) space and time being individually variant in a complementary fashion but taken together they are invariant.   We saw this in the chapters on special relativity and Minkowski space-time.

With the oscillation theory we have a way to explain the physical process that causes the Lorentz-Fitzgerald contraction.   If the oscillation theory is correct then there is a preferred frame of reference, namely the bulk, which by its strong gravity is effectively tied to the rest of the universe.   The universe as a whole, as in Mach's principle, is the preferred reference frame.

As a quantum particle with mass goes at a higher velocity on the TeV brane it is more difficult for the particle to oscillate between the TeV brane and the bulk so its oscillation slows down. The oscillation becomes more difficult because when the particle is going at a higher constant velocity on the TeV brane it has to cover more distance in the bulk between oscillations so it has more exposure to the energy of the bulk and gains mass.   This slows the oscillation of the particle. Along the axis of motion there may be a drag effect as the quantum particles struggle to oscillate between the TeV brane and bulk.   This would cause the quantum particles to bunch up along the axis of motion in the TeV brane squeezing particles together and crowding out particles oscillating back from the bulk.   This creates the Lorentz-Fitzgerald contraction for a stationary observer in the TeV brane reference frame.   The drag effect would not occur perpendicular to the axis of motion so the width of an object would remain unchanged.

What is time?   The oscillation of quantum particles between the TeV brane and the bulk is, I believe, the origin of what we perceive as time.   The slowing of time in special relativity for an astronaut on a rocket departing from earth at high speed and then returning can be explained, as we saw in the Lorentz-Fitzgerald length contraction, by the slower oscillation rate of his quantum particles.   For the duration of the astronauts high speed journey all his constituent quantum particles are oscillating slower so the cumulative effect at the end of his journey is that he has aged less than people left behind on earth.   This would be the physical explanation for time dilation.

The mass increase of quantum particles moving at high speed is also explained by the oscillation theory as we saw in the Lorentz-Fitzgerald contraction.   Quantum particles with mass that are moving at higher velocity have to cover more distance in the bulk during each oscillation thus exposing them to more of the bulk's energy and as a consequence they gain mass.   Note that in our earlier discussion on UFO's eliminating inertial effects I didn't mention that relativistic mass increase was due to oscillation of quantum particles into the bulk.   So just as the strong gravity field of the UFO is needed to replace the 'Higgs' induced mass, this field is also needed to substitute for relativistic mass increase as well.   Most of the mass of protons and neutrons in atoms comes from these relativistic effects.

So with the oscillation theory along with the M-theory inspired RS1 we have a way to explain at the quantum level the key effects of special relativity; length contraction, time dilation and mass increase.   Einstein was correct that the "ether" does not exist as a medium for light waves. Electromagnetic radiation can propagate just fine in a vacuum as determined from Maxwell's equations.   But the oscillation of quantum particles between the TeV brane and the bulk opens up a whole new way to understand modern physics.   Although there is no ether there is a preferred frame of reference if RS1 and the oscillation idea are found to be valid.

Why the Speed of Light is Constant To All Observers
Massless particles have no drag effect from the bulk when they oscillate between the TeV brane and bulk.   Consequently their velocity is constant at the speed of light, at least in a vacuum.   The speed of light is just the maximum possible speed of any massless particle.

All observers no matter what their inertial motion measure the speed of light the same since they and their measuring instruments experience the Lorentz-Fitzgerald contraction via the oscillation theory that exactly compensates for differences between their speed and that of a light beam.   The oscillation of quantum particles between the TeV brane and the bulk is the origin of the preferred reference frame.

Time Dilation in a Strong Gravity Field

In explaining how UFOs minimize or eliminate inertia I used strong gravity fields to slow or stop the oscillations of quantum particles into the bulk which is the source of their inertia. But since I also theorized that our experience of time arises from the oscillation of quantum particles then the slowing of these oscillations in a strong gravity field is the physical cause of gravity induced time dilation as predicted by general relativity.

Higgs

Although particle accelerators have been operating in the energy range where the Higgs particle(s) is expected to be found there has been only fleeting evidence of its existence. Martinus Veltman, a Nobel Laureate, in his 2003 book "Facts and Mysteries in Elementary Particle Physics" notes that some theorists do not believe the Higgs particle even exists. Instead these physicists suspect that the Higgs function somehow mimics a considerably more complex reality that incorporates gravitation in a fundamental way. The oscillation theory does exactly this since it involves the huge number of gravitons thought to be localized in the bulk. The masses of the fundamental particles in the oscillation theory derive from the particle's oscillation into the RS1 warped geometry of the bulk. The oscillation theory along with RS1 provides an alternative explanation for the Higgs function.

Equivalence Principle

At the heart of the theory of general relativity is the equivalence principle which equates gravitational mass to inertial mass. Inertia is defined as the resistance of an object to change in motion, i.e. acceleration. Gravity is sometimes referred to as an acceleration field. If RS1 (based on some version of M-theory) and the oscillation theory are found to be valid then inertia would result from the oscillation of fundamental particles between the bulk and the TeV brane as explained previously. Likewise gravitational mass would also result from the oscillation of fundamental particles between the bulk and the TeV brane including mass derived from relativistic effects. The equivalence of gravitational mass and inertial mass is a consequence of both types of masses sharing a common origin.

Virtual Particles

A possible explanation for virtual particles is that they are a manifestation of particles that are momentarily in the bulk during part of their oscillation cycle. Since time is at or near a standstill deep in the bulk this means that these particles can be everywhere in the bulk at once during half their cycle. A single particle is momentarily smeared out everywhere at once deep in the bulk. If even these "ghost" particles are themselves subject to oscillation then they may manifest on the TeV brane for part of a cycle as what we call virtual particles.

There are no doubt other areas where the proposed UFO technology could be applied to physics on earth. But our primary objective in this book was to unravel the technology behind UFOs and if the proposals presented here are verified then we will have achieved our goal.

In the next chapter we'll do a little speculation on what we humans can expect in the future once we develop earth based UFO technology.

What The Future Holds For Earth Based UFO Technology
Chapter 4 in Section III

The implications of developing UFO technology on our planet are extraordinarily exciting. Some of the possible developments that I think we can expect are fusion power, vastly improved transportation on planet earth, easier and more rapid exploration of our own solar system, quantum leaps in our understanding of the universe and the possibility of locating and exploring other solar systems.

Scientists have been struggling to create fusion power reactors on earth for many decades. So far there has been only limited success with current methods using electromagnetic plasma confinement or lasers for impulse based fusion techniques. But if we can generate immensely strong localized gravity fields then this opens up a whole new approach to achieve controlled nuclear fusion that in effect duplicates how stars fuse nuclei.

The UFO technology we've discussed, as we replicate it here on earth, will allow for incredibly fast travel to any point on planet earth. UFOs have been clocked at over 12,000 MPH in our atmosphere. Besides just raw speed, UFO technology that overcomes inertial forces will permit rapid point to point travel on earth without having to tolerate time consuming low speed approaches and departures from transportation centers. Between high speeds and eliminating inertial effects you can imagine traveling across the continental United States from coast to coast in a mere 15 minutes! Barely time for passengers to be served a bag of peanuts and a drink! By controlling gravity and inertial forces passenger comfort will be assured. The size of the craft will no longer be limited by aerodynamic considerations so leg room and seat width will be spacious. Finally!

Our ability to explore our own solar system will be vastly improved. Whether we opt to travel to other bodies in our solar system by staying on the TeV brane or popping into the bulk the end result will still be vastly easier than using conventional rocket technology. Traveling in our solar system with rockets requires long time periods, large amounts of dangerously volatile fuels, exposure to solar radiation (shunted away into the graviton sink with UFO technology), and time consuming missions of years duration to reach the outer planets. In contrast travel in the bulk or on the TeV brane utilizing the strong gravity/graviton sink technology will permit rapid trips to even the furthest planets in our solar system. Using just the TeV brane might mean a trip to Mars of a few days or maybe a week. But utilizing the bulk we could imagine a trip to Mars in just minutes. With UFO technology the size of the craft is not a significant issue. This means astronauts will enjoy a spacious and comfortable environment. Also by controlling gravity within the craft astronauts will not have to contend with the health problems associated with weightlessness.

With UFO technology our understanding of the universe should increase rapidly. Working with strong gravity fields will undoubtedly reveal important data on the origin of the Big Bang. Easy access to space will allow for construction of space stations and space telescopes of tremendous size.

As we master the technology of accessing and navigating the bulk we will acquire the ability to travel to other solar systems. The early exploration of the bulk will most likely be done with unmanned probes to avoid endangering astronauts. But after the technology of utilizing the bulk has been perfected we will then be able to send astronauts to explore other solar systems. This will be one of the most fascinating aspects of developing UFO technology. However it is a reasonable question to ask how other advanced civilizations will react to our earth civilizations new ability to potentially visit their home planets.   In fact I think we already have some clues as to how they are likely to respond to the possibility of our civilization visiting their home planet.   We'll address this important issue of contact with extraterrestrial civilizations in the next and final chapter.

Contact With Alien Civilizations
Chapter 5 in Section III

A frequently asked about the UFO phenomenon is that if it does represent alien visitors to earth then why don't they make contact?   A related question is why do UFOs seem to carefully avoid being too closely observed by humans?   The answer to these questions may surprise many people but I think it will help explain the often puzzling behavior of UFOs.

So why don't the aliens operating UFOs make contact with our earth civilization?   Abduction stories and conspiracy theories aside there does not seem to be any credible evidence that these alien civilizations have made any attempt to establish contact with our civilization.   Furthermore it is reasonable to assume, given the vast number of sun-like stars in the universe, that we are being visited by not one but many different alien civilizations.   This last observation on the number of alien civilizations visiting earth, I believe, is an important clue to their behavior.

My guess is that there is a federation of advanced civilizations and that they maintain a policy of not contacting or interfering with developing civilizations.   This is why the likely multiple civilizations visiting earth all seem to act in a similar manner.   Their probable definition of a developing civilization is one that has not yet achieved the ability for interstellar travel.   However once a developing civilization acquires the technology to travel to other solar systems via the bulk then I think at that point the emerging civilization is introduced to the federation of advanced civilizations.   This is when contact is finally made.

People are often puzzled by the furtive behavior of UFOs, especially in close up sightings. Typically UFOs will speed away as humans attempt to get closer to these machines.   Why? The answer may be surprising.   The federation of advanced civilizations may know from experience that too close observation of their craft can result in the developing civilization prematurely deducing the technology behind these machines.

So why does this federation wait so long before contact is made?   I think the reason is that developing civilizations can pose a potential threat to more advanced civilizations.   My belief is that all civilizations evolve at roughly similar rates in terms of their technology, social and political systems.   While people might have their doubts I think virtually all developing civilizations become more civilized as time goes on.   I further suspect that the improvement in the nature of planetary civilizations, each consisting of many nation states, is due to the spread of democratic concepts on those planets.   We can see that process occurring on our own planet.   Even dictatorships here on planet earth often label themselves as the "Democratic People's Republic" or the "People's Democracy" etc.   It's clear that among intelligent beings the ideas of self government and individualism is natural and probably unavoidable over time.   Also as communications technology develops it is more difficult for a one party government to control information as we have seen here on earth with the collapse of the Soviet Empire, the appearance of democracy protests in the Iranian theocracy and elewhere in the Middle East.   Competition of ideas and competition in sources of information, i.e. freedom of speech, has a civilizing effect.   To be sure there will be setbacks in the spread of democracy on any planet but in the end I think it is an inevitable process.   The fact that democracies on earth rarely get into armed conflicts with other democracies suggests that this form of government, messy as it may be, makes a civilization more mature and its behavior more restrained.

If there is something like a "Star Trek" prime directive its basic principle will probably be to give developing civilizations the maximum possible time to reach the level of a mature and responsible society before they are finally introduced to the federation.   Contact with the federation most likely occurs when there are indications that a new civilization has begun to access the bulk.

Benefits of Membership in the Federation

Despite the postulated federation of advanced civilizations attempting to conceal UFO technology from developing societies, I'd suspect that many begin to figure it out anyway.   This book itself, if these conjectures prove correct, may be an example of a developing civilization deducing UFO technology from trends in earth based physics and careful observations of UFO characteristics and performance.   If so then we ourselves may soon be taking decisive steps toward fully understanding and implementing UFO technology.   The apparent increase in UFO reports during the latter half of the twentieth century and continuing into the twenty first century may be an indication that the federation knows our technology is reaching the point where our earth civilization is not too far from accessing the bulk.   We humans may find ourselves becoming part of a vast federation of other advanced civilizations in the not too distant future.

Becoming part of such a federation should bring many interesting benefits along with the anticipated responsibilities.   But some other possibilities could include access to a comprehensive history of planet earth going back perhaps millions or even billions of years as recorded by federation members from prior visitations.   Needless to say a visual and audio record of the early history of the earth would be of immense interest.   Imagine experiencing audio/visuals of the earliest life forms on our planet, witnessing the age of the dinosaurs, the evolution of hominids, the ice ages, the building of the Great Wall of China, classical Greek and Roman civilizations in their heyday, etc.   Another area of great interest would be learning more about the origin of the universe and its ultimate fate.   It is likely we will be treated to a fascinating learning experience and over time making significant contributions to federation archives ourselves.

Appendix A - Explaining Maxwell's Laws

To get a better understanding of Maxwell's equations its necessary to have some familiarity with the vector calculus used in their derivation. We'll do a brief review of ordinary calculus first and then check out vector calculus. The goal is to just get an idea of the concepts, understand the symbols and take the mystery out of science. That's all. And not to worry…there won't be any tests!

Basic calculus just involves measuring ***rates of change***. When you drive your car at constant speed, that can be thought of as the ***first derivative*** (in calculus lingo) as you take note of your speed (distance divided by time). When you accelerate your car, that can be viewed as the ***second derivative*** as you observe the rate at which you increase (or decrease) your speed. There! That's pretty much what basic calculus is all about.

In the above example let's assume the car was going on a flat road in a straight line. Then you can use the letter "**x**" to represent the distance you travel on the road and the letter "**t**" can represent the time. In calculus they use the expression **dx/dt** to represent your speed at a particular moment. The **dx** just means "difference" in distance between two points during the associated time period **dt**. In the car example this is the first derivative. Speed is often measured in miles per hour so you could be going at a constant speed of say 30 MPH, 40 MPH, 50 MPH, etc.

Note also that since the car is going in a straight line that you are only talking about one spatial dimension, the **x** dimension.

When it comes to acceleration the calculus expression for the car example is this:

$$\frac{d^2x}{dt^2}$$

This represents the acceleration of the car at any instant and is called the second derivative.

Using simple graphs we can see how the first and second derivatives are used to calculate the speed and acceleration of a car at any instant. The first graph below gives the position **x** of the car at various times **t**. We'll use the following simple formula to describe the position of the car at any time:

$$x = t^3$$ We can call this the ***position formula***.

The first derivative gives you the instantaneous speed at any point on the graph shown on the next page. Mathematicians have come up with rules for determining a derivative of any equation (better them than me!). One of the most common rules is called the ***power rule*** which works like this:

If you have a math expression of the form:

$$at^n$$

Position Graph

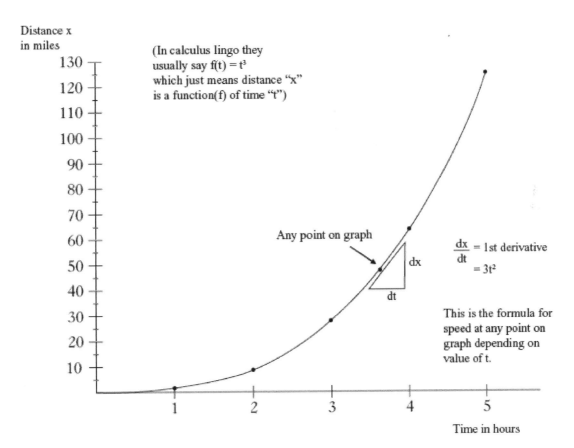

Distance x in miles

(In calculus lingo they usually say $f(t) = t^3$ which just means distance "x" is a function(f) of time "t")

Any point on graph

dx

dt

$\dfrac{dx}{dt}$ = 1st derivative

$= 3t^2$

This is the formula for speed at any point on graph depending on value of t.

Time in hours

$x = t^3$

Time	Distance
1 hour	1
2 hour	8
3 hour	27
4 hour	64
5 hour	125

Where "**a**" and "**n**" are real numbers and "**t**" is the variable (ie. you put in various values for **t**) then the derivative of the function is:

$$(a \cdot n)t^{n-1}$$

Applying this rule to our equation $x = t^3$ where **a** is assumed to be the number "1" (in front of the t^3) and **n=3** we get this:

$$(1 \cdot 3)t^{(3-1)} = 3t^2$$

This first derivative measures the change in distance over the change in time at any point on the graph and represents the speed (velocity) at that instant.   This is **dx/dt**.

A line tangent (parallel) to the position graph at any point can be used as the hypoteneuse of a right triangle with the vertical leg (**dx**) representing change in distance and the horizontal leg (**dt**) representing change in time.   In calculus theory this **dx/dt** ratio is taken to a mathematical "limit" with numerator and denominator made smaller and smaller in tandem which represents the speed (or velocity) at that point on the graph.

To determine the speed at hour 2 just use t=2 in the formula for the first derivative like this:

$$\frac{dx}{dt} = 3t^2 = 3 \cdot 2^2 = 3 \cdot 4 = 12mph$$

We can use the above formula for the car's speed to build the ***velocity graph*** for the car. We'll show speed on the vertical axis and time, as before, on the horizontal axis.   The velocity graph for the car is shown on the next page.

Now if we calculate the 2nd derivative of the position formula (or 1st derivative of the speed formula above) we get acceleration:

The 1st derivative of $x = t^3$ (position formula) was:

Speed (velocity) $\dfrac{dx}{dt} = 3t^2$

The 2nd derivative is:

Acceleration $\dfrac{d^2x}{dt^2} = 6t$ (Just go back and apply the power rule to $3t^2$)

Velocity Graph

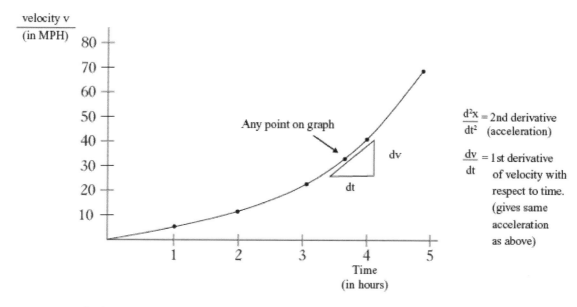

1st derivative = speed = $3t^2$

t (time in hours)	Speed (in MPH)
1	3
2	12
3	27
4	48
5	75

Note:

Mathematicians actually use a variety of symbols to represent 1ˢᵗ and 2ⁿᵈ derivatives. The position formula could be written as $f(t) = t^3$ *instead of as* $x = t^3$. *The* $f(t)$ *just means that the distance x is a function time t. When you use this notation for an equation then you show the first derivative like this:* $f(t)'$. *The accent mark is called prime. For second derivative you show it like this:* $f(t)''$. *This is called double prime. Another way to show derivatives with the* $x = t^3$ *notation is as follows. First derivative is this:* $\dot{x}$. *Second derivative is* $\ddot{x}$. *You can see the single or double dot above the x. Since I use different notations for derivatives in the book I thought I'd better show the different styles.*

So with the second derivative of the position formula we get the car's acceleration. Or we can get the same result by taking the first derivative of the speed formula that we calculated.

$$(ie. \frac{dx}{dt} = 3t^2).$$ Pretty neat, huh?

To determine the acceleration at any point on the velocity graph just insert the time value "t" at that point into the 2nd derivative of the position formula, For $t = 2$ we get:

$$\frac{d^2 x}{dt^2} = 6t = 6 \cdot 2 = 12 \text{ miles per hour per hour (acceleration)}$$

Anyway, as you can see, calculus is pretty easy! In fact it's amazing what you can do with it.
I need to also mention there is another major branch of calculus called *integration*. Integration can be thought of as the inverse of differentiation. In the same way differentiation is the inverse of integration. Just as square roots are the inverses of squares. And visa versa. In our car example how would you calculate how far the car went say between zero hour and time equals 2 hours? It turns out that the area under the curve of the speed graph for that time period will give you the answer! This is integration. Mathematicians have figured out easy rules for determining areas and volumes for lots of different graphs. We won't go into details on those rules but suffice to say the concept is similar to what we did on differentiation.

But for now remember that so far, with the basic differential calculus we have used here, only one spatial dimension, the x dimension representing the distance the car went in a straight line, has been considered. However when you are describing electric and magnetic fields you are working with three spatial dimensions instead of just the one **x** dimension in our car example. This is the situation Maxwell faced. With fields you need a coordinate system with 3 axis. Usually the letters **x**, **y** and **z** are used. Since a field at any point can have both direction and magnitude you need a way to express these factors. Mathematicians have created *vectors* for just this purpose. A vector can be used to describe both the strength and direction of a field at any and all points in the field. A single point in a field using a vector is illustrated on the next page.

The length of the vector **A** can be determined simply by using the three component numbers **a**, **b** and **c** along the **x**, **y** and **z** axis that determine the location of the tip of the vector. Using the Pythagorean theorem in 3 dimensions you can calculate the length of the vector which is its magnitude. The direction of the vector is a function of the numeric values for the components **a**, **b** and **c** on the 3 axis (you can see that by imagining in your mind how the arrow would move about if the numerical quantities **a**, **b** and **c** changed relative to each other and in relation to the **x**, **y** or **z** axis they lie on). If this vector represents a point in an electric field then the length of the vector represents the strength of the electric field at that point (i.e. the base of the vector arrow…called its origin). The vector direction is the direction in which the electric field is pointing. Notice that the letter **A** representing the vector is in bold type, this indicates it is a vector.

Combining calculus concepts with vector concepts we get the branch of math called vector calculus. To get the change in the length and/or direction of the vector at any point in a field (like an electric field) you just calculate the derivative of the vector function. The derivative of a vector

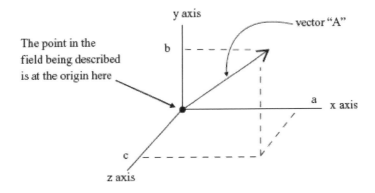

A Single Point in a Field Described by a Vector

function can be another vector function or sometimes a scalar (i.e. just a magnitude with no direction but it does have a position or location).   Just like we had an equation describing the position of the car (in terms of time **t** as the only independent variable) in our basic calculus example, you can imagine that each of the three legs of the vector along the x, y and z axis each has a separate equation describing how its length changes over time, with time as the only variable in each component (leg) equation.   To get the derivative of this vector function you just calculate the derivative of each of the three component legs.   (We'll do an example to make this easy to see).

But it is also possible to have a vector where each of its component legs has more than one variable.   As an example we can imagine a vector where time is not involved but instead there are three variables labeled **x**, **y** and **z**.   Each vector component leg would have its own separate equation with all three variables.   For example the equation along the x axis leg might be:

component leg on **x** axis$= 2x^2 yz^3$.   Oh boy!   How do you calculate the derivative of a function with three variables?!   It's easy.   Just calculate the derivative of the portion involving

the variable x and leave the rest alone.   So the 1st derivative of $2x^2$ is just $4x$ (using

power rule).   The complete derivative of the component leg for the x axis is then: $4xyz^3$.

You do the same process for the other two component legs and then you have a new vector function (derivative) that tells you how that vector changes based on its three variables.   Technically speaking when mathematicians do a derivative of a function with multiple variables but they focus on one variable at a time it is called *partial differentiation*.

If you label the three functions (equations) making up the x, y and z axis components of the vector **A** as P, Q and R we can see what the overall vector function looks like.   The general mathematical form to describe a vector like this is shown below.

Vector $A = Pi + Qj + Rk$

Where **i**, **k** and **j** represent unit vectors on the x, y and z axis.   This just tells you that the P, Q and R functions are to be applied along the x, y and z axis.   So if the P function (along the x axis) is the same as our example above it would look like this:

$P = 2x^2 yz^3$.   The first derivative of this function is: $P' = 4xyz^3$.   (as we saw above).   This 1st derivative tells you how the component vector magnitude (length) along the x axis varies based on the 3 variables x, y and z.   You do the same process for the Q and R functions to determine how those component legs vary according to the x, y and z variables.

Now lets see how we can use vector calculus in Maxwell's laws.   In electromagnetic theory a static electric field is described as *diverging* from a point source electric charge.   This divergence looks like this in equation form:

$$\text{Divergence } A = \frac{\partial P}{\partial x} + \frac{\partial Q}{\partial y} + \frac{\partial R}{\partial z}$$

(Note that for these derivatives we use ∂P instead of dP.   The ∂ is used to indicate that this is partial differentiation.)

But where did this divergence equation come from?   It actually started out as two separate pieces that are joined in a vector operation called the *dot product*.   In abbreviated form the divergence equation showing its two component parts looks like this:

Divergence $A = \nabla \cdot A$

The upside down triangle is called "*del*" and its use means you are using a "*differential operator*".   The del or differential operator is used in vector math on three dimensional vectors to indicate that the differentiation you are performing is partial differentiation.   The del operator looks like this:

$$\nabla = \frac{\partial}{\partial x} i + \frac{\partial}{\partial y} j + \frac{\partial}{\partial z} z$$

The unit vectors (**i**, **j** and **k**) indicate that this partial derivative is to operate on a function that describes the vector components along the x, y and z axis respectively.   The first vector component

218   Solving the UFO Enigma

operator takes a derivative with respect to x only.   The second with respect to y only.   Etc. Notice that there are no (symbolic) vector component leg functions (like P, Q and R) being operated on in the operator!   This del operator is pretty useless by itself!   But that's where our vector **A** comes into the picture.

Our vector function **A** looks like this as we saw above:

Vector $A = Pi + Qj + Rk$

Now we can see what this "dot product" means in vector calculus.   We saw that divergence of vector **A** is abbreviated like this:

Divergence $A = \nabla \cdot A$ (read divergence vector **A** equals del **_dot_** vector **A**)

Substituting our del operator and vector **A** into the divergence formula above we get:

Divergence

$$A = \left(\frac{\partial}{\partial x} i + \frac{\partial}{\partial y} j + \frac{\partial}{\partial z} k \right) \cdot (Pi + Qj + Rk) = \frac{\partial P}{\partial x} + \frac{\partial Q}{\partial y} + \frac{\partial R}{\partial z}$$

Or: Divergence $A = \dfrac{\partial P}{\partial x} + \dfrac{\partial Q}{\partial y} + \dfrac{\partial R}{\partial z}$ (as we saw originally above)

What the "dot product" does is multiply the corresponding components for each of the three axis times each other.   So the partial derivative operator with respect to **x** times the vector component leg P**i** gives us the partial derivative of the function P with respect to **x** on the far right. But what happened to the unit vector "**i**"?   It turns out that in vector rules multiplying **i** times **i** equals one (1).   Ditto for the unit vectors **j** and **k**.   Because the unit vectors disappear in this vector operation the final result is a scalar…ie. it gives the strength of the field at any position but not its direction.

The use of the dot product indicates you are looking at the divergence of a field.

In English this equation just says the divergence (or spreading out) of the electric field, which is represented by vector **A** at any point in the field, is the result of the three partial derivative functions that comprise the components of vector **A**.

This is not meant to be a rigorous definition of calculus and vector calculus but just an overview to see the general concept.   It's also important to see what some of these unusual looking symbols mean to take the mystery out of them.   With this background we can now briefly explain Maxwell's four fundamental equations that explain electromagnetism.

Equation 1

$$\nabla \cdot E = \frac{\rho}{\varepsilon_0}$$

This equation just describes a static electric field generated by a stationary electric charge.   A fixed electric charge produces an electric field that diverges outward in all directions.   In words the equation says that the divergence $(\nabla \cdot)$ (del followed by a dot) of the electric field (E) (in terms of 3 dimensional vectors labeled E) equals the charge density (ρ - Greek letter "rho") divided by **permittivity** (ε_0 - Greek letter "epsilon" with the number zero "0" as a subscript).

The divergence of the electric field is the amount of electric field or lines of force in a region of space.   The divergence is equated to the total electric charge in that region of space divided by the permittivity constant.   Permittivity relates to the fact that different materials have different abilities to transmit an electric field.   In this case the subscript "**0**" (zero) indicates that ε_0 is the permittivity constant for the vacuum of space.

The illustration below shows the divergence of an electric field around a point charge.

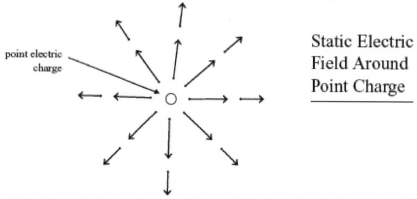

point electric charge

**Static Electric
Field Around
Point Charge**

Notice that vectors for points further away from the charge are shorter than for points closer in. The shorter outer vectors indicate the field strength is weaker at that location.   Also the vectors in the illustration should be thought of as in 3 dimensions not just the 2 dimensions shown.

Maxwell, inspired by the scientist Michael Faraday, was using vector calculus to first describe the whole electric field surrounding a charge and then analyzing it in terms of its pieces or vectors.

His first equation, $\nabla \cdot E = \dfrac{\rho}{\varepsilon_0}$ describing the divergence (or spreading out) of the electric field from a point charge tells you the amount of change in the strength of the electric field as you get further away from the point charge into infinity.

Equation 2

$$\nabla \cdot B = 0$$

This equation just describes a static magnetic field created by a magnet.   In words the equation says the divergence $\left(\nabla \cdot\right)$ of the magnetic field (**B**, magnetic vector) equals zero (0). How come the divergence of the magnetic field equals zero?   Because unlike the static electric field which spreads out in all directions into infinity, the magnetic field "flows" (although doesn't actually move) from the north pole of the magnet to the south pole.   Its net divergence is zero since it comes out of the north pole and disappears back into the south pole.   The magnetic vector field is illustrated below:

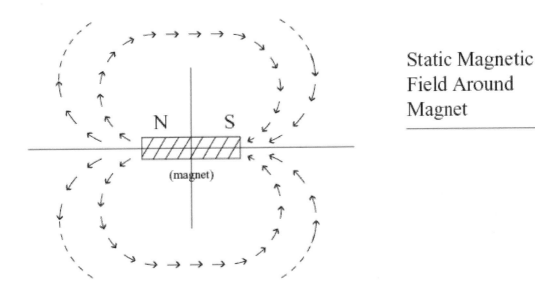

Static Magnetic
Field Around
Magnet

Notice that when the magnetic field vectors are further away from the magnet they are shorter which indicates the magnetic field is weaker at greater distances from the magnet.

Equation 3

$$\nabla X E = -\frac{\partial B}{\partial t}$$

This equation just says that the **curl** (a mathematical operation symbolized by ∇X i.e. del followed by a cross) of the electric field is equal to the negative value of the rate of change of the magnetic field.   More simply this equation just says that a changing magnetic field creates a curled electric field which then pushes electric charges (like free electrons in a wire) around in a circle.   The curlier the electric field the more charges (current) travels in the circle.

Notice the magnetic field B is a vector which means the term on the right side of the equation describes the magnetic field in three dimensions.   That's why you see that this derivative of the magnetic field uses partial derivatives.   Also notice that this equation involves change over time (in the denominator), so time is a factor here.

If you have a loop of wire and you move a magnet through the loop you will induce a circular (or curled) electric field flowing through the wire.   This is illustrated below:

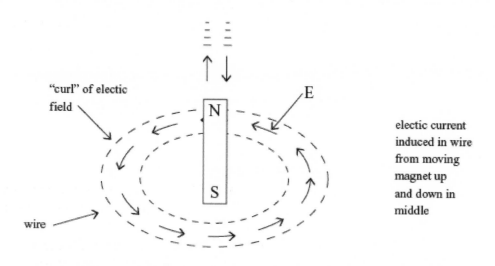

"curl" of electic field

E

electic current induced in wire from moving magnet up and down in middle

wire

Illustration of Electic Field Curl

As the magnet gets further away the induced electric field current gets weaker.   But as the magnet gets closer to the wire loop the electric field current gets stronger.

By moving the magnet in and out of the loop you can generate a pulsating electric current. This is the basis of electrical generators that power our modern civilization.

Equation 4

$$\nabla XB = \mu_0 J + \mu_0 \varepsilon_0 \frac{\partial E}{\partial t}$$

This equation says a curled magnetic field is produced by an electric current (the term $\mu_0 J$)

and/or a changing electric field (the term $\mu_0 \varepsilon_0 \dfrac{\partial E}{\partial t}$).   The first term just means that if you

have an electric current (represented by the vector **J**) going through a wire it will create a curled magnetic field around it.   The second term says that a changing electric field will also create a curled magnetic field around it that varies in strength.   For example if you put an electric charge on the end of a stick and wave the the end of the stick in front of a magnetometer (instrument used to measure a magnetic field) you will see the magnetometer register a (curled) magnetic field that changes in strength as the electric charge gets closer and further away.

Notice the two constants in equation 4 too, ie. μ_0 and ε_0.   We already saw that ε_0 is the permittivity constant associated with the ability of different materials (metal, air, vacuum, etc) to transmit an electric field.   With the zero (0) subscript the ε_0 is the permittivity constant for a vacuum.   The symbol μ (the Greek letter "mu" for m) is the *permeability* of a material, ie. its ability to transmit a magnetic field.   With the zero (0) subscript the μ_0 is the permeability of the vacuum of space.

And that's it!   These are Maxwell's four fundamental equations of electromagnetism.   In the process of describing Maxwell's famous equations we've also introduced basic calculus and vector calculus concepts, some of the mathematics that is at the heart of modern science.

But Maxwell did not stop here.   He did something else that set the stage for the next great advance in science.   If you look at equation 3 you see that a changing magnetic field creates a changing electric field.   Now look at equation 4.   If you assume there is no electric current J you

are left with just the second term ($\mu_0 \varepsilon_0 \dfrac{\partial E}{\partial t}$) on the right side.   This term says that a

changing electric field creates a changing magnetic field.   By combining equation 3 with this modified version of equation 4 Maxwell discovered something amazing.   A changing magnetic field creates a changing electric field.   And a changing electric field creates a changing magnetic field.   Maxwell soon realized that this created a self sustaining electromagnetic wave which is the source of visible light as well as other types of radiation like X-rays, infrared, ultraviolet, etc.   He also calculated that his equations required this electromagnetic radiation to travel at a constant speed of about 186,000 miles per second.   The two constants for the electric field permittivity in a vacuum and the magnetic field permeability in a vacuum are the factors that determine the speed of light in a vacuum.   It would take nearly half a century before the full significance of the constant speed of light would be understood and usher in a revolution in our understanding of nature with Einstein's theory of special relativity.

Appendix B – Energy Scales in Physics

Length and Energy Scales of Physics Theories

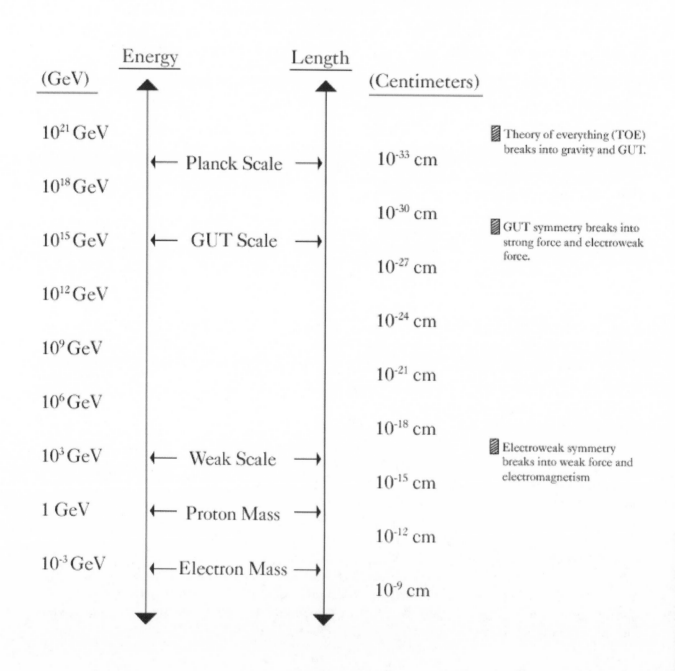

Electron Volt Energy Scales

1 eV	=	1 eV	1 eV
1 KeV	=	1,000 eV	10^3 eV
1 MeV	=	one million eV	10^6 eV
1 GeV	=	one billion eV	10^9 eV
1 TeV	=	one trillion eV	10^{12} eV

Common Conversions and Energy Scales in Physics

10^{19} GeV	=	$10^{19} \cdot 10^9$ eV	=	10^{28} eV
10^{16} TeV	=	$10^{16} \cdot 10^{12}$ eV	=	10^{28} eV

So: 10^{19} GeV = 10^{16} TeV

Planck scale	=	10^{19} GeV	=	10^{16} TeV
GUT scale	=	10^{15} GeV	=	10^{12} TeV
Weak scale	=	10^3 GeV	=	1 TeV
Electron mass	=	10^{-3} GeV	=	1 MeV

Bibliography

Books:

Advanced Aerial Devices Reported During the Korean War, Richard F. Haines. Los Altos, California: LDA Press, 1990.

Aliens From Space: The Real Story of Unidentified Flying Objects, Major Donald E. Keyhoe. New York: Signet Books, 1973.

Atom: Journey Across the Subatomic Cosmos, Isaac Asimov. New York, Penguin Group, 1992.

Above Top Secret: The Worldwide UFO Cover-Up, Timothy Good, New York: William Morrow and Co., 1988.

Beyond Einstein: The Cosmic Quest for the Theory of the Universe, Dr. Michio Kaku and Jennifer Trainer. New York: Bantam Books, 1987.

Black Holes and Baby Universes: and Other Essays, Stephen Hawking. New York: Bantam Books, 1993.

Black Holes & Time Warps: Einstein's Outrageous Legacy, Kip Thorne. New York: W.W. Norton and Company, 1995.

The Charm of Physics, Sheldon L. Glashow. New York: Simon and Schuster, 1991.

Collider: the Search for the World's Smallest Particles, Paul Halpern. Hoboken, New Jersey: John Wiley & Sons, Inc., 2010.

Dreams of a Final Theory: The Search for the Fundamental Laws of Nature, Steven Weinberg. New York: Pantheon Books, a division of Random House, Inc., 1992.

Einstein's Heroes: Imagining the World Through the Language of Mathematics, Robyn Arianrhod. New York: Oxford University Press, 2005.

Exploring the Physics of the Unknown Universe, Milo Wolff. Manhattan Beach, California: Technotran Press, 1990.

Facts and Mysteries in Elementary Particle Physics, Martinus Veltman. River Edge, New Jersey: World Scientific Publishing Co. Pte. Ltd., 2003.

Flying Saucers From Outer Space, Major Donald E. Keyhoe. New York: Henry Holt and Company, 1953.

Flying Saucers: Here and Now!, Frank Edwards. New York: Bantam Books, 1967.

Flying Saucers: The Startling Evidence of the Invasion from Outer Space, Coral E. Lorenzen. New York: Signet Books, 1966.

The God Particle: If the Universe is the Answer, What is the Question?; Leon Lederman with Dick Teresi. New York: Houghton Mifflin Company, 1994.

The Great Beyond: Higher Dimensions, Parallel Universes, and the Extraordinary Search for a Theory of Everything: Paul Halpern. Hoboken, New Jersey: John Wiley & Sons, Inc., 2004.

The Great Design: Particles, Fields and Creation, Robert K. Adair. New York: Oxford University Press, 1987.

Hiding in the Mirror: The Mysterious Allure of Extra Dimensions, From Plato to String Theory and Beyond, Lawrence M. Krauss. New York: Viking Penguin, 2005.

The Hynek UFO Report, Dr. J. Allen Hynek. New York: Dell Publishing Co., 1977.

Hyperspace: A Scientific Odyssey Through Parallel Universes, Time Warps, and the 10th Dimension, Michio Kaku. New York: Oxford University Press, 1994.

Introducing Relativity, Bruce Bassett and Ralph Edney. USA: Totem Books, 2002.

In Search of Schrodinger's Cat: Quantum Physics and Reality, John Gribbin. New York: Bantam Books, 1984.

In Search of the Ultimate Building Blocks, Gerard 't Hooft. Cambridge, United Kingdom: Cambridge University Press, 1997.

The Lonely Sea: The Autobiography of Francis Chichester, Francis Chichester. New York: Coward-McCann, inc.: 1964.

The Matter Myth: Dramatic Discoveries That Challenge Our Understanding of Physical Reality, Paul Davies and John Gribbin. New York: Simon and Schuster, 1992.

Night Siege: The Hudson Valley UFO Sightings, Dr. J. Allen Hynek and Philip J. Imbrogno with Bob Pratt. New York: Ballantine Books, 1987.

The Official Guide to UFOs, The Editors of Science and Mechanics Magazine. United States: Davis Publications, Inc., 1968.

Out of This World: Colliding Universes, Branes, Strings, and Other Wild Ideas of Modern Physics; Stephen Webb. New York: Praxis Publishing Ltd., Copernicus Books, 2004.

The Phoenix Lights: A Skeptic's Discovery That We Are Not Alone, Lynne D. Kitei, M.D. Charlottesville, Virginia: Hampton Roads Publishing Co., Inc., 2010.

Q is for Quantun: An Encyclopedia of Particle Physics, John Gribbin. New York: The Free Press, A Division of Simon & Schuster Inc., 1998.

The Quantum Frontier: The Large Hadron Collider, Don Lincoln. Baltimore, Maryland: The John Hopkins University Press, 2009.

The Quantum World: J. C. Polkinghorne. Princeton, New Jersey: Princeton University Press, 1989.

RelativityDemystified: a Self Teaching Guide, David McMahon. New York: McGraw-Hill, 2006.

Schrodinger's Kittens and the Search for Reality: Solving the Quantum Mysteries, John Gribbin. Boston: Little, Brown and Company, 1995.

Science at the Edge: Conversations with the Leading Scientific Thinkers of Today, Edited by John Brockman. New York: Sterling Publishing Co., 2008.

The Search for Superstrings, Symmetry and the Theory of Everything, John Gribbin. Boston: Little, Brown & Company, 2000.

Search for a Supertheory: From Atoms to Superstrings, Barry Parker. New York: Plenum Publishing Corporation, 1987.

Situation Red, The UFO Seige, Leonard H. Stringfield. Fawcett Crest Books, 1977.

Strange Skies: Pilot Encounters with UFOs, Jerome Clark. New York: Kensington Publishing Corp., 2003.

Superstrings: And the Search for the Theory of Everything, F. David Peat. Chicago: Contemporary Books, 1988.

Symmetry: And the Beautiful Universe, Leon M. Lederman and Christopher T. Hill. New York: Prometheus Books, 2004.

Time, Space and Things; B.K. Ridley. Cambridge, England: Press Syndicate of the University of Cambridge, Canto Edition, 1995.

Trigonometry Demystified: a Self Teaching Guide, Stan Gibilisco. New York: McGraw-Hill, 2003.

Uncertainty: Einstein, Heisenberg, Bohr, and the Struggle for the Soul of Science, David Lindley. New York: Random House, Inc., 2008.

The UFO Book: Encyclopedia of the Extraterrestrial, Jerome Clark. Detroit, Michigan: Visible Ink Press, 1997.

UFO: The Complete Sightings, Peter Brookesmith. New York: Barnes and Noble, 1995.

The UFO Experience: A Scientific Inquiry, J. Allen Hynek. New York: Ballantine Books, 1972.

UFOs: A Pictorial History from Antiquity to the Present, David C. Knight. New York: McGraw-Hill Book Company, 1979.

UFOs: A Scientific Debate, edited by Carl Sagan and Thornton Page. New York: Barnes and Noble, 1996.

UFOs: The Secret History, Michael Hesemann. New York: Marlowe & Company, 1998.

Understanding Physics: Three Volumes in One: Motion, Sound and Heat; Light, Magnetism and Electricity; The Electron, Proton and Neutron; Isaac Asimov. New York: Barnes and Noble, 1993.

The Universe in a Nutshell, Stephen Hawking. New York: Bantam Books a division of Random House, Inc., 2001.

The Unknown Universe: the Origin of the Universe, Quantum Gravity, Wormholes, and Other Things Science Still Can't Explain, Richard Hammond. Franklin Lakes, New Jersey: The Career Press, Inc., 2008.

Unveiling the Edge of Time: Black Holes, White Holes, Wormholes; John Gribbon. New York: Crown Trade Paperbacks, 1992.

Vector Analysis, Schaum's Outline Series, Murray R. Spiegel. New York: McGraw-Hill Book Company, 1959.

Warped Passages – Unraveling The Mysteries of the Universe's Hidden Dimensions, Lisa Randall. New York: HarperCollins Publishers, 2005.

What is Quantum Mechanics?: A Physics Adventure, Transnational College of LEX.   Boston: Language Research Foundation, 1997.

Magazines:

Scientific American, published by Scientific American, Inc.; selected issues.

International UFO Reporter, published by the Center for U.F.O. Studies, selected issues.

Organizations and Web Sites:

Best UFO Resources.   Hyper.net/ufo.html.

J. Allen Hynek Center for U.F.O Studies (CUFOS), Chicago, Illinois.   CUFOS.org.

Fund for UFO Research.   FUFOR.com.

Mutual UFO Network.   MUFON.com.

National Aviation Reporting Center on Anomalous Phenomena.   NARCAP.org.

National Investigations Committee on Aerial Phenomena.   NICAP.org.

National UFO Examiner with Roger Marsh.   Examiner.com/ufo-in-national.

National UFO Reporting Center (NUFORC).   NUFORC.org.

TheOzFiles.com website.

The Phoenix Lights website of Dr. Lynne Kitei.   ThePhoenixLights.com.

Project Blue Book, code name for the United States Air Force study of U.F.O's, formerly based at Wright Patterson AFB in Ohio, USA.   Via National Archives.

UFO Casebook with Billy Booth.   UFOCasebook.com.

UFODigest.com website.

Extraterrestrial Contact – Scientific Study of the UFO Phenomenon and the Search for Extraterrestial Life.   UFOEvidence.org.

National Press Club Conference of November 12, 2007.   Testimony from Captain Jean-Charles Duboc on his crew sighting of a UFO on January 28, 1994.

Physics Archives

"Black Hole Particle Emission in Higher Dimensional Spacetimes" by Dr. Vitor Carloso, Dr. Marco Caxaglia and Dr. Leonardo Gualtieri.   arXiv:hep-th/0512002v3 28 Apr 2006

Index

41135559R00142

Made in the USA
Middletown, DE
04 March 2017